Time and Society

Time and Society

Warren D. TenHouten

State University of New York Press

Published by
State University of New York Press, Albany

© 2005 State University of New York

All rights reserved

Printed in the United States of America

No part of this book may be used or reproduced in any manner whatsoever without written permission. No part of this book may be stored in a retrieval system or transmitted in any form or by any means including electronic, electrostatic, magnetic tape, mechanical, photocopying, recording, or otherwise without the prior permission in writing of the publisher.

For information, address State University of New York Press, 90 State Street, Suite 700, Albany, NY 12207

Production by Michael Haggett
Marketing by Michael Campochiaro

Library of Congress Cataloging-in-Publication Data

TenHouten, Warren D.
Time and society / by Warren D. TenHouten
    p. cm.
Includes bibliographical references and indexes
ISBN 0-7914-6433-4 (hardcover: alk. paper)
1. Time—Sociological aspects. I. Title.

10 9 8 7 6 5 4 3 2 1

For Maisie and Kevin Cavanagh

Contents

Preface		ix
Acknowledgment		xiii
Chapter 1	Introduction	1
Chapter 2	A Case Study of the Australian Aborigines	11
Chapter 3	Patterned-Cyclical Time Consciousness	25
Chapter 4	Patterned-Cyclical Time Consciousness, Continued	41
Chapter 5	Ordinary-Linear Time Consciousness	57
Chapter 6	Patterned-Cyclical and Ordinary-Linear Time and the Two Sides of the Brain	69
Chapter 7	Immediate-Participatory and Episodic-Futural Time and the Brain	83
Chapter 8	The Two and the Four, and Possibly More: Social Duality and the Four Elementary Forms of Sociality	95
Chapter 9	Natural and Rational Experiences of Time	111
Chapter 10	Communal Sharing and Patterned-Cyclical Time Consciousness	121
Chapter 11	Equality Matching and Immediate-Participatory Time Consciousness	135
Chapter 12	Authority Ranking and Episodic-Futural Time Consciousness	157
Chapter 13	Market Pricing and Ordinary-Linear Time Consciousness	169
Chapter 14	Text and Temporality	187
Chapter 15	An Empirical Test of the Theory	195
Chapter 16	Discussion	207

Appendix	217
Notes	221
References	223
Name Index	245
Subject Index	251

Preface

> There are philosophical topics other
> than time. Some of them engage our attention
> for extended periods. But sooner or later each
> is sucked down into the vortex of the one problem
> that is truly permanent: time ... the focus and horizon
> for all thought and experience.
> —David Wood, *The Deconstruction of Time*

My interest in time consciousness in its relation to the social world developed out of a study of Australian Aboriginal cognition, which I undertook with the encouragement of Lex Grey, a native New Zealander, who had worked with the indigenous Maoris of New Zealand to develop an early family education program. With the aid of the Van Leer foundation of the Hague, the Netherlands, he brought Maori leaders to Australia, the result being a highly successful program in twenty-four rural communities in New South Wales (NSW), the NSW Aboriginal Family Education Centres Federation (AFEC). Grey, and the AFEC Aborigines, in 1984, invited me to Sydney to meet with them about conducting a study of the thinking style of Aboriginal children. They were motivated by the conviction that Aborigines have a distinctive mentality, that this mentality should be taken into account in any project devoted to preparing children mentally for the beginning of formal education, and that the positive cognitive resources Aborigines possess should be taken notice of, and considered a positive resource, by the educational system. To this end, with the involvement of the president of the AFEC, Maisie Cavanagh, and Kevin Cavanagh, along with other participating Aborigines too numerous to mention, an initial study of cross-cultural differences in cognition was carried out.

The Australian Aborigines can be considered exemplars of oral, indigenous culture; the Euro-Australians, of modern, Western culture. Any comparison of these two cultures is confounded insofar as they have different socioeconomic levels, differential participation in a culture of poverty, different race-group membership, and so forth. To discover which factors might be contributing to observed cognitive differences, a questionnaire was designed for parents. To avoid superimposition of my own, Western, academic categories in such an instrument, it was decided that a number of life-historical interviews would be conducted. Analysis of these interviews was helpful in discovering the ways in which Aborigines categorize their world. A close reading of the interviews revealed that Aborigines, in comparison to controls, tended not to use words that refer to cause and effect, number, sequence, quantity, and clock time, but to use more words pertaining to space, surfaces, patterns, and the cycles of nature and of social life.

To explore these findings, the corpus of life-historical interviews was greatly expanded, to 658, and a lexical-level content analysis of the transcripts was pursued. A decision was made to begin analysis with a consideration of time orientation, because of my long-standing interest in time and temporality. This led to a nearly exhaustive study of the literature on Aboriginal time consciousness, which is without doubt more extensive and fully developed than is the study of time in any other nonmodern culture, bar none. Thus, rather than review time consciousness in several or many oral, indigenous cultures and extend the study to include "peasant" or "traditional" cultures as well, it was decided to focus on a single case study. Also required was a study of the vast literature on time and temporality in general as it has developed in philosophy and science. In the study of the Aborigines' sense of time, together with reading this vast literature on time and temporality, six things gradually became apparent and guided the inquiry: (1) The distinction between linear time and cyclical time is important to any understanding of time consciousness and must be included in any serious theory of time and society; (2) The concepts of linear and cyclical time are *both* poorly defined in the literature, so they must be given rigorous, grounded definitions, a task undertaken in this book; (3) There exist subjective experiences of primordial temporality that are neither cyclical nor linear, that do not fit into these poorly defined categories; (4) These additional experiences of time are an "episodic-futural" temporality based on conation and intentionality and an experience of the knife-edged present, the "immediate-participatory" experience of the lived mind and body; (5) A theory of time and society requires a model not only of time but also of society, which has been developed by the author over two decades; and (6) It is not enough to merely explicate a theory of time and society, for it is essential that any theory of a general nature must not only be testable in principle but actually put to an empirical test so its truth or falsity can, at least in a preliminary way, be established.

The word "theory" has been used loosely in my professional field, sociology, and in the study of time and temporality in general. The conceptualization presented in this book, however, *is* a theory in a rigorous sense. It has propositions—six in number, elementary and derived concepts, and hypotheses deduced from the propositions which are empirically testable. Using Peter Roget's ([1852] 1977) remarkable classification of the English language, I have developed a simple methodology for testing the hypotheses, which is based on a quantitative lexical-level method of content analysis. It will be shown that the hypotheses of the study are consistent with the word-usage measures of Roget folk concepts applied to the life-historical interviews, which provide direct, empirical evidence in support of the theory.

My research in the field was indirectly supported by a grant from the University of California's Pacific Rim Research Program, which enabled me to spend a semester as a visiting professor of sociology at the University of New South Wales. Data acquisition and data processing have been greatly facilitated by a series of grants from the University of California at Los Angeles (UCLA) Academic Senate. Computer resources were provided by the UCLA Office of Academic Computing, with special thanks to statisticians Mel Widowski and Xiao Chen. I am grateful for the support, encouragement, and facilitation and for the direct involvement in the project that has been provided by the AFEC. Thanks also to Stuart Reed of the Battye Library in Perth, Janette Blanco of the Macksville, NSW, Technical and Further Education Program, Tom and Jean Hay, and sociologist Dr. Steven D'Alton, formerly of the University of New South Wales. Again, my deepest appreciation to Lex Grey, Maisie and Kevin Cavanagh and other members of their extended family, and numerous other Aborigines who have been supportive and helpful, especially my friend, the late Wally LeBrocq at Old Burnt Bridge, Kempsey, NSW. I thank my wife, the sociologist Maria Gritsch, whose insight and critical commentary were most helpful in developing and refining my arguments. I thank Dr. Charles Kaplan at Rijksuniversiteit Limburg, Maastricht, the Netherlands, for his valuable commentary on the materials pertaining to Heidegger and to other theoretical issues and problems. Political scientist William Kitchen stimulated my interest in content analysis as a way to study cognitive structure, and philosopher Andreas Koch read and criticized the entire manuscript. Finally, the life-history study has been worked on by over 1,000 undergraduate students and several graduate students at UCLA, who gave generously of their time and energy: of these students, special thanks go to Daniel Cardenas, Jocelyn Su, Jocelyn Gibson, and An Phung.

Acknowledgment

Acknowledgment is made for Parts of this book that have previously appeared elsewhere. Some of the material in chapters 1 and 2 originally appeared as "On Durkheim's Notions of Time and Mind in Australian Aboriginal Cosmology and Social Life," in *The Living Legacy of Marx, Durkheim & Weber*, vol. 2, ed. R. Altschuler. New York: Gordian Knot Books, 2000. Some of the material in chapters 6 and 7 appeared as "Primordial Temporality, the Self, and the Brain," *Journal of Social and Evolutionary Systems* 20:3: 253–79 (1997). Some of the material in chapters 7 and 8 appeared as "The Four Elementary Forms of Sociality, Their Biological Bases, and the Implications for Affect and Cognition, in *Advances in Human Ecology*, vol. 8, ed. L. Freese, pp. 253–84, JAI Press: 1999. Some of the material in chapter 14 appeared as "Text and Temporality: Patterned-Cyclical and Ordinary-Linear Forms of Time-Consciousness, Inferred from a Corpus of Australian Aboriginal and Euro-Australian Life-Historical Interviews," *Symbolic Interaction* 22:2: 122–37 (1999), reprinted with permission of the University of California Press. The data presented in Chapter 15 appeared in "Time and Society: A Cross-Cultural Study," *Free Inquiry in Creative Sociology* 32:21–34 (2004).

Chapter 1

Introduction

What distinguishes human beings from the rest of the animal world? "Intelligence" has been a key word in the answer to this philosophical and scientific question. One approach to this problem is psychometric testing, the effort to measure and compare individual and group differences in the quality and effectiveness of mental functioning. An alternative approach is to study the use of important *categories* of human reasoning. In every culture the major notions used to conceptualize reality are what Aristotle called "categories of the understanding," including the seen and the unseen, the present and the absent, the changing and the changeless, and matter and spirit. The cosmologies and epistemologies of modern societies, capitalist and socialist alike, are based above all else on two such cosmic categories—time and space. Perhaps the most basic aspect of a person's mentality is the way in which the major categories of life are lived. Of these categories, the person's experiences of time and space are arguably the most fundamental. The essential categories of reason that dominate our intellectual life, Durkheim contended, "correspond to the most universal properties of things. They are like the solid frame which encloses all thought … for it seems that we cannot think of objects that are not in time and space, which have no number, etc. … They are like the framework of the intelligence" ([1912] 1965: 21–22). While time and space are inseparable, time retains a certain priority over space in modern societies (Whitrow 1989: vi–vii). This priority, however, is nearly reversed among members of indigenous, oral cultures, such as the Australian Aborigines, where considerations of time are easily and commonly recast in terms of space.

In the sociological, anthropological, and related literatures on time and temporality, several empirical generalizations, on the surface, seem to be well established. The literatures of this subject, spanning more than a century, reveal a general consensus that modern, Western people have a "linear" perspective on time, and that they tend to be "future oriented." Additionally, it is known in the study of time in sociology (Adam 1990) and in social anthropology (Gell 1996) that members of "primitive," preliterate, and pretechnological cultures have a "holistic," "cyclical" view of time that is a synoptic, all-at-once gestalt (e.g., Barnes 1974) and also have a "present orientation." There also is evidence in these social-scientific literatures that these cross-cultural differences in time consciousness are related to cultural evolution. For example, it is well established that in hunting-and-gathering, preliterate, and tribal societies, communal social relations are given great emphasis, and that, in many instances, these societies show a remarkable insistence on a principle of social equality (Itani 1988). And, in modern, Western societies, and far beyond, social relations, of course, *include* communal and equality-based informal social relations but in addition are organized in a more formal way, with an emphasis on economic activity and on a competitive social hierarchy. In spite of these generalizations, there remains much to do in the study of time and social organization.

Daunting problems remain regarding the concepts' definitions. The linear-cyclical distinction is taken for granted by many "temporal dualists" but rejected out of hand by those who reject social duality theory. The notion of "cyclical" time is fundamentally a metaphor that might or might not include "alternating" or "flattened" time, and it remains badly in need of a clear definition (see Gell 1996: 30–36, 84–85, 91–92). Linear time, assumed by many to be self-explanatory in its meaning, is based on the thinnest of rationales to be the "opposite" of cyclical time. A second distinction is between present or near-present orientation and future orientation, which presents its own problems of clear conceptualization. Beyond these four concepts that are often, but not always, contrasted as two *pairs* of temporal orientation, of linear versus cyclical and of present-past versus futural, there are, of course, nearly endless ways to conceptualize temporal experience. But it should be considered premature to give up on these broad notions of time awareness, because they are of fundamental importance and, it will be argued here, can be considered culturally universal cognitive structures. In this book, attention will be confined to just these four kinds of time consciousness—cyclical, linear, present oriented, and futural—with the full awareness that other kinds of time consciousness are of great importance and descriptive value.

A seven-part definition of "cyclical" time, which will be called "patterned cyclical," has been developed inductively by the author through a protracted study of the "primitive" civilization with *the* most elaborated study of its time orientation, the Australian Aborigines. It will be found that when these seven aspects of patterned, cyclical time are, as a methodological operation, turned into their opposites or near opposites, the result is a full description of our ordinary, linear notion of time, based as it is on clocks, calendars, schedules, and timetables. Neither of these two kinds of time consciousness, having seven-part definitions, are dualities, and both can be regarded and measured as continuous variables, so that any cross-cultural differences that might exist can be considered a matter not of kind but of degree. An effort also is needed, and made here, to clearly define present-oriented and future-oriented kinds of time consciousness. Moreover, these resulting definitions can and will be *criterion validated* by showing that they are important aspects of the four most general kinds of information processing that contemporary cognitive neuroscience has been able to associate with the functioning of the organ of all thought and symbolic reason, the human brain.

There also are serious conceptual difficulties in relating social organization to time consciousness. For example, it has been established in social anthropology and sociology that members of primitive culture tend in their social lives to emphasize *communal* social relationships and, equally important, to hold to a principle of social *equality* (and, of course, to have many other important features as well). But how, in these cultures, are the social relations and forms of time consciousness related? Might their cyclicity of time orientation be related to communal social relationships, or to the effort to establish social equality, or to both, or to their interactions? And might their present orientation result from communal relationships, or from equality in social interactions, or from both, or from their interactions? Remarkably, this *specification* of just which aspects of social organization contribute to kinds of time orientation has not been addressed.

An analogous situation obtains in the study of time in modern society. For modern, Western societies, there is a vast literature linking industrialization, capitalism, urbanization, postmodernization, and other social macroprocesses to both linearity and futurality. These social macroprocesses involve *social relations* based on economics and politics, referred to by Fiske (1991) as "market-pricing" and "authority-ranking" social relations, respectively. Once again, there are, of course, innumerable other ways to conceptualize participation in modern life, but here attention will be based on relations of money and power. There certainly would seem to be a consensus that market-pricing and authority-ranking social relations influence the modern tendency toward linearity and futurality, but again it is far from clear *how* these social and cognitive variables are linked. Does our involvement in economic relations make us linear thinkers, or do relations of social power, or is it a combination? And what makes us futural in our temporal orientation? Economic social relations or power-based relations, or both? Once again, theory and data are needed. In the theory presented and tested in this book, four kinds of time consciousness will be paired, in four propositions, to the *positive* experiences of four social relations. In one of these pairings, however, which sees the linearity of time consciousness influenced by market-oriented, economic social relations, we will be in for a surprise with respect to culture and the *valence* of involvement in market-pricing social relations.

Two other theoretical possibilities, which to the best of my knowledge have never been raised in the study of time and societal development, are: (1) that cyclicity and present orientation, which are obviously complementary and closely linked, might together form an *emergent* level of time consciousness, and (2) that linear and futural time orientations, which also are complementary and closely linked, might similarly interact to form an *emergent* level of time consciousness. The theory presented in this book explores these possibilities explicitly.

A parallel argument can, and will, be made about social organization. We have referred to four aspects of social relations that also seem to pair in a natural way. First, communal-sharing and market-pricing social relations, involving the principles of communion and agency, respectively, can be seen as being opposite in their meanings. And second, in instances of hierarchical, authority-based relations being set aside, or suspended, the result is an opposite situation, a *conditional equality*.

It also is important to consider the valences of these social relations. For example, communal-sharing social relations are positive as we enjoy the company of companions, community members, and friends, but are negative as relations become hostile, abusive, and destructive. Thus we will, in the process of theory construction, confine our attention to four social relations, but we will need eight variables to do so. It will be proposed (1) communal-sharing and equality-based social relations, which are complementary in any society and together are apt to be fundamental to primitive societies, together in their interactions result in a higher level of social organization, the basis of informal sociality, which can be called "hedonic community," and that (2) economic and political social relations, which also are complementary and prevail in modern societies, result in a higher level of social organization, the basis of formal sociality, which Chance

(1988) calls "agonic" society. This model, it will be shown, finds its basis and its criterion validation in primate and human ethology.

THE THEORY

The theory of social relations and time consciousness presented in this book is both general and complex. It is general in that it conceptualizes kinds of time consciousness as aspects of the most general modes of information processing known to be associated with the human mind and brain. It causally links these kinds of time consciousness to basic problems of life that have engendered the most fundamental kinds of social relations. And the theory is complex in that it is based on three levels of analysis—the mental, the social, and the biological, so the cognitive model of time consciousness and the sociological model of social relations are both shown to have a biological basis and an evolutionary history.

1. The *cognitive* model of time consciousness posits the existence of four elementary, irreducible forms of time consciousness: (1) the ongoing experience of the present is termed *immediate-participatory*; (2) immersion in the patterns and cycles of nature and social life, often referred to as "qualitative" or "cyclical" time, will be termed *patterned-cyclical*; (3) our experience of what Martin Heidegger ([1927] 1962) called "ordinary" time, involving the measure of motion with clocks, calendars, and schedules, is termed *ordinary-linear*; and (4) efforts to plan for and bring about the future is termed *episodic-futural*. These four kinds of time, once explained, will then be grouped in pairs into two of the highest-level kinds of temporal experience: first, the unity of our simultaneous involvement in the present and in the cycles and patterns of life engenders a *natural* time experience; and second, the unity of our ordinary-linear and episodic-futural experiences of time engenders a *rational* time experience (the topic of chapter 9). While the four kinds of temporal experience will be seen as cultural universals, the same is not true of natural and rational time experience, which might or might not emerge as cognitive structures in the mind of any individual human being. An argument will be made that the rational and natural kinds of temporal experience are not only different but are opposites, because ordinary-linear time and patterned-cyclical time are opposites, and immediate-participatory time and episodic-futural time—involving temporal compression and temporal stretching, respectively—also are opposites.
2. The second level of theory is *biological*. The most modal functioning of the brain and nervous system, it will be proposed, provides the

infrastructure for the four elementary forms of temporality. The neuroscientific literature on brain functioning and the experience of time includes some rather abstract, mathematical models of a putative "internal clock" in the brain (e.g., Treisman et al. 1990; Rammsayer and Vogel 1992). These models, however, do not contradict the simple fact that there is no evidence for a "clock" in the brain providing a direct measure of elapsed duration; human beings have not developed the capacity to know "what time it is," as measured by a clock, beyond close approximations. Insofar as time consciousness involves the processing of information on a very general level, it is a natural step to postulate that the proposed four elementary forms of temporality might be expressions of the most general modes of information processing of the human brain.

Luria (1973) and Pribram (1981) have identified a fundamental polarity associated with the cognitive functioning of the frontal and posterior lobes of the brain. The frontal lobes are the command-and-control center of the brain and are the biological infrastructure of cared-about plans and intentions. Pribram (1981) proposes that the frontal lobes, working closely with limbic and other structures, generate a mode of information processing that he calls "episodic." The posterior lobes of the brain—parietal, occipital, and temporal—are in contrast responsible for sense perception and the construction of the present moment, a kind of information processing that Pribram calls "participatory." The immediate-participatory mode of primordial temporality, our mindful immersion in the present, is based on the primary information processing of the three posterior lobes of the brain.

A secondary but still important polarity in brain organization is found in the division of labor between posterior regions of the left and right cerebral hemispheres, which are specialized for two kinds of information processing that have been interpreted as opposites by numerous philosophers, psychologists, neuroscientists, and others. Bogen (1969) refers to the left and right hemispheres, especially in the right-handed adult with a reasonably normal brain, as usually but not always specialized for "propositional" and "appositional" modes of information processing, respectively. Levy-Agresti and Sperry (1968) refer to these same lateralized modes of information processing as "logical-analytic" and "gestalt-synthetic." Bogen makes a critical point when he writes of "[w]hat may well be the most important distinction between the left and right hemisphere modes [of information processing], ... the extent to which a linear concept of time participates in the [left hemisphere's] ordering of thought" (1977: 141, emphasis deleted). It should be cautioned that all neuroscientists do not accept the inference of two kinds of information processing from the experimental literature, taking this to be an inflationist view of cerebral lateralization theory carried out by scholars with a commitment to a generous view of the boundaries of human consciousness (Efron 1990; Corballis 1991; but cf. TenHouten 1992a). It is proposed that the ordinary-linear mode of time consciousness is an aspect of the logical-analytic, *serial* mode of information processing of the left side of the brain, and patterned-cyclical time consciousness of the gestalt-synthetic, global, and *simultaneous* information processing of the right side of the brain.

3. The third level of analysis is *sociological*. A model claiming the existence of exactly four elementary forms of sociality will be presented. Florian Znaniecki in 1934 proposed that every human cultural system possesses a hierarchy of one supreme dominant element and unspecified numbers of dominant and subdominant elements. These dominant elements of a cultural system he posited to be culturally universal which, he suggested, would require that they have a biological basis and an evolutionary history. He offered no substantive interpretation of these elementary forms of sociality. The task of specifying such a model, carried out in this book, begins with a consideration of social duality theory. The idea that there are two kinds of society is an ancient one, finding current expression in philosophy, psychology, sociology, anthropology, and primate ethology. The distinction made is between formal and informal social organization, which parallels Tönnies' ([1887] 2000) still-useful distinction between the folk *Gemeinschaft* (community) and the exchange *Gesellschaft* (society), and their associated mentalities of Natural Will and Rational Will, in whose honor the natural-time/ rational-time distinction is introduced. Max Scheler's (1926) elitism and careless scholarship have made him a nearly forgotten figure in the history of social theory, but he left behind a highly cogent idea. He conceptualized four elementary forms of sociality, paired under two larger principles: (1) kinds of being with one another; and (2), the kind and rank of values in whose direction the social members see with one another. Scheler conceptually unpacked the first, informal level of society, seeing that it contained two elements, identity and life community. Formal society, in his view, was based on two kinds of social relations, rank and value, corresponding to the institutional domains of politics and economics.

A remarkably similar conceptualization was developed by Robert Plutchik ([1962] 1991) in his psychoevolutionary classification of emotions. He proposes that there are, for all animal species, four fundamental, existential problems of life: identity, temporality (reproduction), hierarchy, and territoriality. He said that the negative and positive experiences of these existential problems led to eight prototypical adaptive reactions that define eight primary emotions that together form secondary emotions. There is a close conceptual agreement between Scheler's identity, life community, rank, and value, and Plutchik's identity, temporality, hierarchy, and territoriality. More recently, Fiske ([1991] 1992) has identified four elementary forms of social life, which he terms *equality matching* (EM), *communal sharing* (CS), *authority ranking* (AR), and *market pricing* (MP). Fiske had, without reference to any of the aforementioned scholars, replicated Scheler's notion that there are four basic things we do in the social world, which we can combine in complex ways. As will be explained, there are some serious conceptual problems with Fiske's model, but his terminology is well known and will be used in this book.

An argument will be made that these four functions exist as a double polarity. It will be proposed that equality-matching and authority-ranking are opposite human tendencies, either to make things unequal between people (AR) or to contradict this hierarchy and to create a state of conditional equality (EM). Communal-sharing and market-pricing social relations, it will be further proposed, are a result of opposite relations of the individual to society, with society within the individual in CS and the individual within the society in MP. The social model that emerges, the topic of chapter 8, has a structure isomorphic to the mind's dimensions of information processing in general, and time processing in particular, for it is a dynamically related double polarity in which the elements of agonic society are the opposites of those of hedonic community. It will be shown that these concepts are consistent with recent advances in human and primate ethology.

Chapter 2 explores the history of time in the minds of colonizers and the colonized, and in this context it introduces our case study, the Australian Aborigines, who will play an important role in this book. A study of Aboriginal culture and its concept of time has been used to inductively develop the model of *patterned* and *cyclical* time consciousness. The meaning of the familiar duality of linear and cyclical time is neither rejected nor taken for granted but is instead subjected to a rigorous process of definition. The present conceptualization was realized, inductively, through a case study of Aboriginal time consciousness. The author carried out three years of ethnographic field research while living with Aboriginal extended families. As a result of this fieldwork, together with a study of the relevant literatures, seven closely related aspects of Aboriginal time consciousness were identified, all based on the patterns and cycles of nature and culture (see chapters 3 and 4). It will be shown in chapter 5 that when these seven aspects of cyclical time are turned into their logical opposites, the result is a full description of our ordinary, clock- and calendar-based "linear," one-dimensional time. The biological bases of the four kinds of time consciousness are established next, with chapter 6 devoted to the ordinary-linear/patterned-cyclical opposition with its basis in psychophysiology and cerebral lateralization theory; chapter 7 focuses on the immediate-participatory/episodic-futural polarity with its basis in two of the three highest-order functional units of the brain, the posterior cortical unit for processing and storing information from the outside world and the unit for programming, regulating, and verifying mental activity (Luria 1973: 43). With this preparation, it then becomes possible to postulate the social conditions under which the four kinds of time consciousness are used, discussed in chapters 9–12. At the simplest level, the theory holds that participation in agonic society and involvement with power and resources contribute to a time consciousness that is, in its emphasis, episodic-futural and ordinary-linear, and comprises a rational experience of time; hedonic social relations, involving relations of social equality and inequality, contribute to an immediate-participatory temporality of the sensed world, that together with a patterned-cyclical time consciousness, engender a natural experience of time.

There are refinements of these basic propositions, and these statements are certainly not meant to suggest a one-way causality, with social relations the independent variables and kinds of time consciousness the dependent variables. The situation is, of course, more complex than that, and there is every reason to anticipate reciprocal effects. For example, the process of becoming a stockbroker leads to work experience that will sharpen and intensify time consciousness that is episodic and predictive, oriented to the linear dimensions of time and money. On the other hand, a person who already possesses an effective episodic-linear, rational time consciousness is apt to become a stockbroker.

The theory presented in this book is, from a logical point of view, quite simple and straightforward in its meaning. The propositions of the theory are testable. The theory is tested by a radical cross-cultural comparison, between Australian Aborigines and Euro-Australians. It will be argued, in chapter 14, that life-historical interviews provide the richest possible source of data for the empirical study of time consciousness and temporality. In order to carry out such a test, a lexical-level, quantitative, content-analytic methodology has been developed for indirectly measuring social relations and time variables, based on a remarkable classification of the English language developed by Roget ([1852] 1977) in his *International Thesaurus*. Roget was able to sort individual words into 1,042 "broad classes of ideas," here called "folk concepts," with the uses of words in many of these categories serving as multiple indicators of eight social-relations variables—the negative and positive experiences of identity matching, communal sharing, authority ranking, and market pricing and four time consciousness variables—immediate-participatory, patterned-cyclical, episodic-futural, and ordinary-linear. The data set consists of 658 transcripts of life-historical interviews with Aborigines and Euro-Australians obtained throughout Australia in a wide range of ecological settings. It will be shown that the theory fits the data. The methodology for the study will be explicated in the context of a discussion of text and temporality, the topic of chapter 14. The study, and its results, will be presented in chapter 15, and these results will be discussed in chapter 16, which also will contain a general discussion.

Chapter 2

A Case Study of the Australian Aborigines

In the nineteenth and early twentieth centuries there were numerous investigators of the social origins of time. "Primitive," or more precisely, "preliterate," societies, with their exuberant celebrations based on uneven calendars of ritual and ceremonial life, had gripped the minds of many social thinkers. Kern elegantly summarizes this vision in the following:

> The sociology and anthropology of that age was full of information about primitive societies with their celebration of the periodic processes of life and the movement of heavenly bodies, their vital dependence on seasonal change and the rhythmic activity of plants and animals, their exotic commemoration of ancestral experience, and their cyclic and apocalyptic visions of history. (1983: 19)

It is, Kern adds, no wonder that classical social theorist Émile Durkheim came to believe in the social relativity of time. Durkheim's primary interest, however, was not time and temporality but rather the categorical distinction between the sacred and the profane, which he saw as the fundamental requisite for religion, for dual-symbolic classification systems, and for social differentiation and the division of labor. His inquiry, however, could not have proceeded far without the concepts of time and temporality. In *Primitive Classification* ([1903] 1963), he and Marcel Mauss mentioned only in passing that time is related to social organization. But in his 1912 study of Australian Aboriginal religion, *The Elementary Forms of the Religious Life*, he explored the subject in some detail, making a distinction between "private time" and "time in general," the latter having a social origin.

Following Aristotle's and Kant's notions that the contents of reason are categories, Durkheim saw time as a category. The foundation of the category of time, he argued, is the articulated rhythms and patterns of nature and society. Durkheim wrote that for the Aborigines, "the division into days, weeks, months, and years, etc. corresponds to the periodical recurrence of rites, feasts, and public ceremonies" ([1912] 1965: 23). While Durkheim was mistaken in claiming that traditional Aboriginal cultures have "weeks"—which is but an arbitrary cultural form with no natural referent—he had correctly realized that members of human societies organize their lives in time and establish sociotemporal rhythms. At least part of Durkheim's motivation for studying the Aborigines was his view—shared by most social anthropologists, social evolutionists, and ethnographers of his day—that "primitive" tribes afford what Gellner calls "a kind of time-machine, a peep into our own historical past, ... providing closer evidence about the early links in the Great Series, ... the Great Path to the modern world" (1964: 18).

Durkheim flatly asserted the inferior mentality of the Aborigines, a view based on no real evidence but consistent with his orientalism and involvement in the socio-evolutionary theorizing of his time. The Aborigines, in part as a result of their lack of material possessions and nomadic way of life, were widely regarded as a "protoid form of humanity incapable of civilization, at the very bottom of the evolutionary scale" (see Stanner 1965: 209). Durkheim's choice of the Aborigines as a case study was based on his opinion that they are a society as slightly evolved as possible, representing "the lowest and simplest form of social organization" ([1912] 1965: 80). The social evolutionists implicitly posited an identity of social and biological evolution, so a group with a "savage" social organization also must have savage "minds." The presence of totemism in such a society, closely identifying humans with animals, was seen in and of itself as certain proof of the inferiority of the so-called "savage," who was on this basis isolated from civilized man, this isolation largely taking the form of conquest and dispossession.

The Aborigines' island-continent, Australia, was during this era seized upon as a picture of antiquity and strangeness over the whole range of its phenomena. For the eugenicists, instinct theorists, social evolutionists, and social Darwinists, Australia was a place where progress had stopped, with the Aboriginal population eking out a living in a land that time had forgotten. According to this viewpoint, as aptly summarized (but not embraced) by Chase and von Sturmer, Australia "represented a stage very close ... to that at which man had originated; the past had been miraculously preserved and for those interested in the question of "origins" here was a fleeting opportunity which needed to be grasped before it vanished inevitably under the impact of Western civilization" (1973: 4). Australia thus became a major area of anthropological, biological, psychological, and sociological research. It was seen as a place "out of time" where, according to Sir James Frazer,

> ... many a quaint old-fashioned creation, many an antediluvian oddity, which would long ago have been rudely elbowed out of existence, in more progressive countries, has been suffered to jog quietly along in this preserve of Nature's own, this peaceful garden, where the hands of the dial of time seem to move more slowly than in the noisy bustling world outside. (1961: 1023)

This concept of social evolution, which placed Australian Aborigines at the lower rungs of the ladder of evolutionary progress, attracted the attention of neuroanatomists seeking Aboriginal brains, to measure their size, structure, and level of evolutionary development (e.g., Shellshear 1937). It also lured psychologists eager to measure the "intelligence" of the Australian savage for the gratifying purpose of comparing it to that of their own Caucasian race. Fiske's statement on the Aboriginal mind and brain summarized the pseudo-scientific European assessment of these "Stone Age" people, as he averred:

> If we take into account the creasing of the cerebral surface, the differences between the brain of a Shakespeare and that of an Australian savage would doubtless be fifty times greater than the differences between the Australian's brain and that of an orangutan. In mathematical capacity, this same Australian, whose language contains no words for justice and benevolence, is less remote from dogs and baboons than from a Howard or a Garrison. In progressiveness, too, the difference between the lowest and the highest race of men is no less conspicuous. The Australian is more teachable than the ape, but his limit is nevertheless very quickly reached. (1893: 71–72)

The Aborigine, from this socio-evolutionary perspective, was viewed as having no chance of successfully "adjusting to the New European Order of Australia in the foreseeable future. He was a highly specialized creature delicately adjusted by mechanism adapted to a unique environment" (Chase and von Sturmer 1973, 7). Any disturbance of the relationship between the Aborigine and nature, it was widely believed, could only result in the extinction of the fragile, genetically inferior, and noncompetitive Aborigine (Smyth [1878] 1972: 244). Collier wrote: "Their disappearance was a natural necessity. It came about in obedience to a natural law. It was affected by natural processes, and followed on the lines of the substitution of vegetable and animal species all over the world" (1911: 129–30). Of the Aborigines, J. D. Woods opined:

> ... without a history, they have no past; without a religion, they have no hope; without the habits of forethought and providence, they can have no future. Their doom is sealed, and all that the civilized man can do ... is to take care that their closing hour shall not be hurried on by want, caused by culpable neglect on his part. (in Talpin et al. 1879: xxxviii)

History was to bear out neither Woods's prediction nor his moral advice. The Aborigines were subjected to "culpable neglect" of genocidal intensity (Grayden 1957; Reynolds 1974; Bloomfield [1981] 1988; Hughes 1987; Elder 1988). In the southeast of Australia, the Aboriginal population had, after first contact, been culled by an estimated 96 percent a generation before the first ethnographies of the region were written and published. The Aborigines of Tasmania were virtually exterminated, in part through sport hunting and torture.

The doctrine of Aboriginal inferiority was used as a justification for both the seizure of the Aborigines' lands and resources and their close supervision and control—including virtual confinement on missions, cattle stations, and government reserves. They were collectively rounded up after being dispossessed of much of their lands or taken away from their mothers and placed in "modern" institutions on the ground that they were "half-caste" and should be assimilated into white society. Aborigines, full-blooded and half-caste alike, were quickly pressed into the cash economy, initially without pay and in conditions of a *corveé* labor force not far removed from slavery, where they were apt to be subjected to a "hiding" if they failed to report to work at a specified time—a rude introduction to clock time, or they ran away. Gradually, Aborigines took on a variety of proletarian occupational roles, some degrading and some not. Today they live in towns, suburbs, and urban areas, and many are culturally assimilated, participating in the educational system, the cash economy, and the other institutions of modern, Western Australian culture.

The doctrine of differential intelligence failed to grasp that any valid measure of intelligence must be taken in relation to the environment in which thinking takes place. The early psychometric testing of Aborigines had been carried out in a way that placed them in an unfamiliar situation, one in which many cultural assumptions of the testing situation were not met. This "Intelligence Quotient" testing led to the conclusion that the Aborigine lacked general intelligence, possessed only a "Stone Age" mind, and was not fully modern in evolution (see Gould 1981). Darwinist

Thomas Huxley ([1863] 1906), in 1863, speculated that the Aborigine might well be the "missing link" in the evolution of the human species.

One assumption on which this now discredited body of research rested was that of linearity. Since Broca (1873), it came to be assumed that human races could be ranked on a *linear* IQ scale of mental capability. But this assumption is itself based on a particularly modern, Western way of thinking, of propositional, logical-analytical thinking. Broca's efforts to estimate intelligence from brain size ran afoul of his own data, which showed that Eskimos, Lapps, Malays, Tartars, and several other peoples of the Mongolian type surpassed the most "civilized" peoples of Europe. Rather than abandon his theory, Broca resorted to a tortured logic, arguing that the direct relationship between brain size and intelligence might fail at the upper end of a linear scale but works at the lower end, because small brains belong exclusively to people of low intelligence (see Gould 1981: 86–87).

The European perspective on the Australian Aborigines was that they have long inhabited a remote, an insulated, and a timeless world, without a history of their own, without a religion to explain their own origins, and without any technology to reckon and measure time. In addition, the Europeans held the view that the Aborigines, confronted by the change in environment brought about by their conquest and dispossession, according to Wake, "would no doubt develop new areas of intellect, but this would require considerable time for gradual change, and this, in the face of rapidly increasing European incursion, is precisely what the Aborigines would not have" (1872: 82). The Aborigines, it was held, were running out of time and were heading rapidly toward extinction. The "failure" of Aborigines to adjust as expected to European civilization was attributed to a mental inability to make such a profound transition. The Aborigines, it was implied, possess an insufficient temporality, and they were seen as having an insufficient level of intelligence to involve themselves in the history of "the lucky country," Australia. As Chase and von Sturmer summarize this position: "Some retardatory principle had obviously been at work. In these circumstances attempts to civilize were doomed to failure. The Aborigines had been left behind by History and were locked back in the remote past" (1973: 11).

The Australian Aborigines: History, Culture, and Social Organization

Beginning with the 1770 landing of Captain James Cook at Botany Bay, on the eastern shore of Australia, the British, Irish, and other European

FIGURE 2.1   A well-known photograph of Aboriginal prisoners, taken in the late nineteenth century in Western Australia.

colonists of the island-continent of Australia came from various and sundry backgrounds and with a wide range of agendas. Some came out of scientific curiosity, others to seek fortune, looking for mineral wealth and introducing horses and cattle for pastoral enterprise, and yet others came as adventurers and explorers, and in a unique, never-to-be-repeated experiment, an entire cohort of British prisoners was forcibly transported to Australia. English lawmakers had acted to largely rid their country of its "criminal class" and a putatively empty continent presented itself as a fortuitous destination. Australia became "a cloaca, invisible, its contents filthy and unnamable" (J. Bentham 1812: 7). Bentham described the Anglo-Saxon subject matter of the experimental "thief-colony" as "a peculiarly commodious one; a set of *animae viles*, a sort of excrementious mass, that could be projected and accordingly was purposefully projected as far out of sight as possible" (*ibid.*). An unexplored continent was to become a dystopia, a veritable jail. The space around Australia, the transparent labyrinth of the South Pacific, would come to be seen as a wall 14,000 miles thick. Australia's second wave of colonizers, unlike the first, Aboriginal wave, bore no resemblance whatsoever to Rousseau's image of "natural man moving in moral grace among free social contracts, but man coerced, exiled, deracinated, in chains" (Hughes 1987: 1), as illustrated in figure 2.1. The indigenes of other parts of the Pacific, Hughes added, "especially Tahiti, might seem to confirm Rosseau. But the intellectual patrons of Australia, in its first colonial years, were Hobbes and Sade" (*ibid.*).

The European settlers developed both positive and negative perceptions of the Aborigines. On the positive side, some saw in the Aborigines what

Rosseau and the Romanticists had imagined, the Noble Savage. There exists among contemporary Euro-Australians what might be called the "Noble Savage syndrome," which has, as Tatz (1982: 10) puts it, resulted in an adulation of the "loin-clothed ... idealized type" to the detriment of the "non-traditional" and "detribalized" type. Parallel to this cultural definition is a racial distinction between "full-blooded" and "half-caste" Aborigines. There is in Australia a widespread popular attitude that the only "true" Aborigines are those who are overtly traditional and full-blooded, especially those living in tribes. This reality can be seen in land rights legislation: Aboriginal groups that have been most successful in obtaining their fullest land and human rights have been those who are overtly traditional (Jacobs 1988: 31–32).

But on the negative side, the colonists who were free and in positions of power, officers and gentlemen, in large measure did not view the Aborigines as Noble Savages but rather as ignoble, ignorant, unintelligent, atavistic, animalistic, animistic, and superstitious, and as a less-evolved version of the human, the Stone-Age predecessors of the modern human. The description of Aborigines as "Stone Age" people was metonymic, to make them one with "Early Man." The displacement of Aborigines by settlers was, according to Beckett, seen as "a process that had been prefigured in evolution" (1988: 196). Fusing the principles of cultural and biological evolution, Aborigines came to be seen as anachronisms whose extinction was a natural, if pathetic, fact of life. They belonged to another time and would not be staying long. The colonization of Australia was part and parcel of European imperialism. Annexation of the colonized lands, the outward movement of people and goods, and the expansive ideology of imperialism were "spatial expressions of the active appropriation of the future" (Kern 1983: 92). Writing with solemn responsibility, British Liberal-Imperialist Foreign Minister Lord Rosebery interpreted motivation for colonization in terms of the future:

> It is said that our Empire is already large enough, and does not need extension. That would be true enough if the world were elastic, but unfortunately it is not elastic, and we are engaged at the present moment ... "in pegging out claims for the future." We have to consider not what we want now, but what we shall want in the future.... We have to look forward ... to the future of the race of which we are at present the trustees. (cited in Langer 1935: 78)

The naked Australian savages, with their heathen beliefs and customs, were, as justified by theories of biological determination and imperialism, accorded no social status, so their lives, claims to lands, and culture were accorded little respect; they were all to often treated with systematic intolerance and prejudice. They were ignored and kept out of sight, and vast areas were declared empty territory, as specified by the legal doctrine of *terra nullius*[1] (Rowse 1994: 21). The Aboriginal philosophy, contemplative and symbolist in belief and expression, while a matter of scientific curiosity to anthropologists, was dismissed out of hand on the grounds that it was pantheistic, animistic, heathen, and primitive. The Aborigines saw themselves as intimately linked to natural and to cosmic forces, in a cosmology that was antithetical to European thought. Not understanding of or understood by the European settlers, and not invited to dialogue, the Aborigines found their lands expropriated, their freedom to pursue a nomadic way of life largely taken away, their families intentionally broken up by having their children taken away (Edwards and Read 1989), their culture disapproved of and not respected, the use of their languages proscribed, their bodies violated and brutalized, and their women at risk of being raped with impunity.[2] While remote Aboriginal societies have clung to their ancient culture and way of life, many groups of Aborigines found themselves living on the fringes of Euro-Australian settlements, on missions and ministries, in incision on pastoral lands, in urban slums, and, by a gradual process, in working-to-middle-class suburban and urban communities.

At the time of the European "discovery" of Australia more than two centuries ago, the Aboriginal population of the island-continent of Australia was roughly 1 million persons (Butlin 1983; White and Mulvaney 1987). A comparison of ethnographies of Australian Aborigines (e.g., Warner [1937] 1958; Meggitt 1962; Tonkinson 1978; Myers 1986; Rudder 1983; 1993) suggests that the Aborigines have historically participated in a common, "Aboriginal" way of living. They are the contemporary bearers of an ancient civilization. They have, over tens of thousands of years, refined their hunting-and-gathering mode of economic production (F. G. G. Rose 1987).

Language and dialect areas, in general, define the various Aboriginal societies, or "tribes." These areas, in turn, are defined by environmental conditions and boundaries, especially climate and sources of fresh water (Kirk 1981). Figure 2.2 shows the approximate locations of the tribes of Aborigines that are analyzed in this book. The geographic partitioning of Aborigine tribes is to some extent reificational, because extended families, hunting bands, moieties, clans, and languages have overlapping and indistinct boundaries and experience systematic changes of membership. Much of the ritual life of Aborigines involves social gatherings of multiple tribes and can include individual Aborigines whose identification with a single tribe is problematic. Clans tend to be patrilineal, and women typically marry outside of their clans of birth. Further, some clans within tribes speak

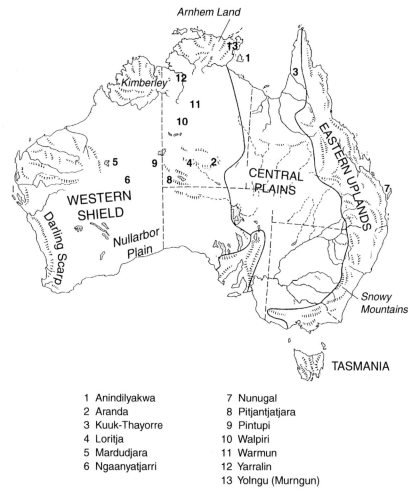

FIGURE 2.2 Approximate location of Australian Aboriginal "tribes," including those analyzed in the text.
Source: Kirk 1987: xvi.

different dialects (and are apt to be polyglots), meaning that in some cases a child's mother and father will speak different dialects (and perhaps languages). Since the time of Captain Cook, the indigenes of Australia have come to develop a Pan-Aboriginal consciousness and a common language, which ironically is English, in fringe settlements, a creole English. Near the end of the eighteenth century, when Europeans, primarily from the British Isles, began to colonize Australia, the number of Aboriginal languages was estimated to range from 150 to 650 (Oates 1975; Wurm 1972), with 250 being a reasonable point estimate (by Yallop 1982: 29).[3] The wide variation in these estimates results not only from a paucity of linguistic research but also from the fact that neighboring tribes have close contact with each other and speak similar languages, making it difficult to determine if one is dealing with dialects of a single language or with separate but closely related languages. Classifications of Aboriginal languages tend to be conservative, counting entire dialect chains as a single language or "family-like language" (Yallop 1982).[4] The lexical diversity of the Australian languages is so wide that the phrase "The Aboriginal word for ..." is apt to have no meaning. This divergence of vocabulary, as Yallop (1982: 38) notes, is in sharp contrast to phonetic and grammatical similarities. For example, the central-desert Arunta (now called the "Aranda") languages share many features of pronunciation and grammar, which set them apart from neighboring languages, however, the languages have quite different vocabularies. With respect to the topic at hand, it is useful to note that Australian Aboriginal languages generally have no verb "to count" and no noun meaning "year."

Maddock's excellent description of Aboriginal society reminds us that "one has ... to add L. R. Hiatt's observation that anthropologists in Australia have moved behind the advancing frontier to A. W. Howitt's observation that the frontier in Australia has been marked with a line of blood" (1972: ix). Swain (1991: 151), summarizing an elaborate literature on the subject, writes that entire communities were murdered or died of introduced diseases, and that many tribes were decimated before being researched or having their very existence documented.

Spiritual Beliefs

Aboriginal culture and society, once discovered, was quickly exploited as a valuable field of opportunity for anthropological and sociological research. Durkheim, in his 1912 *Elementary Forms*, set out to discover the most primordial element of religious life that he postulated would be definitive of religion itself. He realized that his task would be difficult indeed were he to study major theologies of the modern world, with their complex cognitive and symbolic elaborations, including efforts by Descartes and others to establish logical proofs of the existence of God.

Durkheim—along with many other seminal scholars of his day, including Charles Darwin—decided to study the simplest and most "primitive" of all religions, which he claimed were "the crude cults of the [central] Australian tribes," which he asserted were based on "the crudest beliefs" ([1912] 1965: 14, 332). The term *primitive religion*, as used by Durkheim ([1912] 1965: 13), was defined by two criteria: that it be found in a society whose organization is surpassed by no other in simplicity, and that the religion can be explained without reference to elements borrowed from earlier religions. Durkheim's methodological strategy for finding the most elementary forms of religious life consisted in making the Aborigines a case study, which he analyzed using only secondary data. Durkheim never visited Australia and in all probability never met an Aborigine. He reached a still-controversial conclusion, to wit, that the most elementary aspect of religion is neither a God concept nor an avataric figure but rather a dual symbolic classification system that distinguishes between two sides of reality, the sacred and the profane.

His choice of the Aborigines as a primitive culture was a sound one for at least two reasons: first, the Aborigines possess the world's most ancient civilization, based on a hunting-and-gathering mode of economic production, which has continually existed in Australia for an estimated 50,000 to 60,000 years (Roberts, Jones, and Smith 1990); second, the Aborigines' culture is profoundly theocratic. Even in absorbing food, the tribal-living Aborigine partakes in the sacredness of the world, so that living as a human being is itself a religious act (Eliade 1973: 62). Aboriginal totemism is predicated on the belief that all forms of life are interconnected. Human beings and all other sorts of creatures—plants, animals, reptiles, birds, and insects—share the same life essence; none is independent of the other. Totemism specifies a relationship between a person, or a group of persons, and a natural object or spirit being. Totems are of two kinds: the mythical beings connected to clans or local descent groups that are linked to certain sites in that group's territory, and the conception-birth totem (in some societies, the "spirit child"), which enters the mother and animates the child to be within her. Generally believing in reincarnation, the Aboriginal worldview holds that birth represents a movement from the significant, invisible, and temporally prior situation to the present, visible situation. The term *reincarnation*, however, should not be used loosely or taken to have the same meaning that it does in Hinduism. It is not the individual self that is reborn, as the dead are not believed to retain their psychic unity, possibly beyond a limited existence as ghosts. The person's identity is rather conceived of as a fragment of the life potential of a place-imbued Ancestral Being. Life is, above all, seen as an outgrowth of a site, a place on the land, so that when life ends, its spirit essence returns to its place. In the absence of outside influence—which has led the Aborigines to embrace notions such as "The Land of the Dead" and "Heaven"—Aborigines see death as the return of a place-being to place. "The culmination of life," Swain writes, "reaps neither rewards nor punishments for a further life. It is enough that the spirit is restored to its place" (1993: 45). Swain also points out that "wistful regard of Aborigines concerning the possibility of continued life is to be found precisely in those areas of Australia where alien contact has been most extensive and where there has been the transfer of some of the organizational role of place to the body and to time" (1993: 46).

The spiritual beliefs of the Aborigines, in spite of regional and tribal differences, have an overall common cosmology that has been variously referred to by non-Aboriginal anthropologists as their "Dream Time" (Gillen 1896), "Dreaming" (Stanner 1968), and "World Dawn" (Radcliffe-Brown 1945). Aborigines in general prefer the English word "Dreaming" in reference to their cosmology, which is focused on "an epoch beyond which the imagination does not go and which is considered the very beginning of time … [at which time] there were certain beings who were not derived from any others. For this reason, the central Australian Arunta call them the *Altjirangamitjina*, the uncreated ones, those who exist from all eternity" (Durkheim [1912] 1965: 281; see also C. Strehlow 1907: 2).

Aboriginal spirituality has been systematically undermined by the 1951 Commonwealth goal of forcing assimilation and its aftermath, by its denigration as superstition and primitive myth, by having their lands appropriated, by having their native languages discouraged not only at school but in the home as well, by having their ritual life strongly discouraged, and by messianic Christianity in the form of missionaries and their alien message. Aborigines have been encouraged, even forced, to give up their old ways, their *modus vivendi*; losing their land, they have had to nearly give up their nomadic way of life. But even when Aborigines are Christianized or identify with other universal creeds, they have done so in a way that ensures that their older ideas of ubiety also endure. Today Aborigines are apt to describe themselves as a people with "two Laws" (Swain 1991: 165).

In spite of the degradation and oppression that has traumatized and brutalized the Aborigines since their conquest and colonization, the persistence of their culture and cosmology has shown them to be not "weak beings," as Durkheim fantasized, but rather as a remarkably strong and resilient race of people. The Dreaming is a metaphysical expression of primordial notions explaining the birth of the world and showing one's place in this world. The Dreaming is not a physical place with resident spirit-beings but is rather a place of metaphysical repose.

John Rudder (1983, 1993) has carried out a protracted ethnography of the Yolngu (also, Yolŋu or Murngun), the Aboriginal people of north-east to north-central Arnhem Land. The Yolngu is a loosely linked society of

3,000 to 4,000 people, speaking a number of related dialects. Rudder found that the Yolngu speak of their cosmology, in spatial terms, as a two-sided cosmos, with one side subject to transformation and the other unchanging, which they describe as the "outside" reality (*warranul*) and the "inside" reality (*djinawa*), respectively. The outer reality is that of everyday, mundane, and profane experience; the inner reality, the eternal cosmos, the sacred source and destiny of everything. All that exists, according to this cosmogony, derives from a single unchanging source. Experience in this culture emphasizes the continuity and permanence of this eternally present world of nonordinary reality. The Yolngu, and Aborigines in general, do not keep records and do not trace the extended lineage to the Beings who performed the great events of the Dreaming. The link to the Dreaming and the rhythmic events of life are rather conjoined through place.

Dubinskas and Traweek (1984: 28) explain that Walbiri's *Jukurrpa* (Dreaming) is not primarily a "time" but rather a symbol that shapes the experience of the Ancestral realm, this "realm" being both a location and a source of energy. Swain (1993: 24) notes that if we insist on retaining the Dream-root, then *Jukurrpa* would best be translated "Dreaming-Event" (a qualitative not temporal distinction) or "Dream-Place."

The *illud tempus* of the Dreaming, along with its emphasis on space and place, suggests that Aboriginal cosmology gives priority to space over time as the central concept. This stands in sharp contrast to Western civilization, where priority is give to time over space. But this does not mean that time is unimportant to the Aborigines: on the spatial level, the Dreaming links the inner and the outer realities, but on the temporal level, it links the past and the present. There is, indeed, a veritable penetration of the present by the past in Aboriginal temporality, which is nearly definitive of tradition and which finds its most intensity during the performance of rituals. In the next chapters we shall see that past-present fusion is a defining attribute of patterned-cyclical time consciousness.

Chapter 3

Patterned-Cyclical Time Consciousness

This chapter and the next one describe a seven-attribute model of a specific kind of nonlinear, holistic, synoptic time consciousness, to be termed *patterned-cyclical*. These attributes of patterned-cyclical time consciousness were identified through a nearly exhaustive reading of the literature on Australian Aboriginal time consciousness and were retained on the grounds that their *opposites*, or *near opposites*, are interpretable as features of one-dimensional, ordinary, linear time. Not every feature of Aboriginal time consciousness fits this definitional requirement. Most importantly, there is evidence that Aborigines are present oriented, which requires some sort of temporal *compression*. The opposite of temporal compression, a temporal *stretching*, which yields a time consciousness largely determined out of the future (secondarily out of the past), is episodic and futural oriented, not linear. It should be emphasized, at the outset, that while this kind of time was *found* in ethnographic descriptions of a single culture, it is in no way confined to members of this culture. On the contrary, patterned-cyclical time will be seen as universal, existing as a category of understanding the world in all people, regardless of their language, social status, or culture.

Elkin described three aspects of Aboriginal time consciousness, one of which is to be discarded and two of which will be incorporated into the present model. First, in an obviously mistaken claim endorsed by no one else, Elkin posited that for the Aborigines, time "is a series of then and now, the now almost immediately becoming then" ([1938] 1979: 234). This experience of time, as fleeting and flying by, as transitory, will be shown in chapter 6 to be a feature of linear, clock time.

Second, Elkin ([1938] 1979: 232, 234) identified one of seven aspects of Aboriginal time consciousness—a fusion of the past and the present.

And third, Elkin ([1938] 1979: 235) and Stanner (1979) must both be credited with the insight that Aboriginal time consciousness has the attribute of being "cyclical" as opposed to "linear." Elkin's oversimplified, dualistic equation of "Western: Aboriginal :: linear: cyclical," along with the more general formulation, "modern: primitive :: linear: cyclical," has been effectively criticized (Eliade [1954] 1991; McElwain 1988; Chen 1992; Swain 1993; Adam 1990: 127–48). It is untenable to reduce time in contemporary, modern society to the linear and to see in premodern societies only a cyclical timelessness of endlessly recurring sameness. We will see that the putative cyclicity of Aboriginal time consciousness need not be discarded but rather can be placed under a broader and, importantly, different category of temporal experience.

The Aborigines' time consciousness, and patterned-cyclical time consciousness in general will be described as having (at least) seven distinct aspects: it is dualistic, being split into two levels of reality, the sacred inner reality and the profane outer reality (P1); there is in significant ritual interactions, and more generally in tradition, a fusion of past and present (P2). Aboriginal time consciousness will be further described as irregular and heterogeneous (P3), event oriented (P4), cyclical and based on overlapping and interdependent patterns and oscillations (P5), qualitative (P6), and based on the experience of long duration (P7).

Dualistic: Split into Two Levels of Reality

Aboriginal cosmology posits two levels of reality, the sacred, extraordinary, inner reality of the Dreaming and the profane, ordinary, outer reality of everyday life. On this, Durkheim wrote, "The religious life and the profane life cannot co-exist in the same unit of time. It is necessary to assign determined days or periods to the first, from which all profane occupations are excluded" ([1912] 1965: 347). This temporal partition between the sacred and the profane can be clearly seen in tribal-living Aborigines' religious practices. In any religion—Aboriginal or not—there is a division of time into two alternating parts, so the continuity and homogeneity of duration are broken up, made discontinuous, by the theology's calendar, a calendar that concentrates ritual and ceremonial life into periods of time in which the sacred comes to the fore.

Aboriginal sacredness is of the soul, which is a sacred thing opposed to the profane body. In a wide range of "primitive" and even modern societies, the left hand and left side of the body are considered profane, and the right hand and right side of the body are sacred. It is remarkable that Durkheim, who argued that the sacred-profane distinction is *the* fundamental requisite for any and all religions, ignored this crucial fact, and in his *Elementary Forms* ignored his protégé, Robert Hertz ([1909] 1960), who had carried out a brilliant analysis of related phenomena among the Maori, who were known to actively suppress the left hand of children, even binding the left arm to the side in order to control negative spirituality, such as witchcraft, sorcery, and related "left-handed practices" (Coren 1992; Fabbro 1994). There is evidence from early ethnographies in New South Wales that young, initiated women were apt to have the little finger, or the second finger, of the left hand removed (Phillip 1914; Howitt 1825), and that in some tribes a new husband would bite off the wife's first joint of the third finger of the left hand (Cartmill 1955). In the Nunugal tribe at Amity Point, Queensland, there was a custom of cicatrization of the left breast of pregnant women (Welsby 1917). Such practices suggest that Aborigines might well have associated the left hand and left side of the body with the profane and ritually acted to control the profane forces that the left side symbolized and embodied.

Stanner (1979: 24) refers to the Aborigines' cosmology as a sacred, heroic time long ago when humans and nature came to be as they are now. The split inherent in the Aboriginal conceptualization of time is a necessary aspect of the typical Aboriginal worldview within which moral authority is embedded in the Dreaming. Communication with the "other side" is a central preoccupation of Aboriginal spiritual life. The most important communication of all is, of course, the *plan of life* itself. The all-encompassing scope of this life plan, Maddock writes, is shown in "the configurations of the landscape, the events narrated in myths, the acts performed in rites, the codes observed in conduct, and the habits and characteristics of other forms of life" (1972: 110). Communication with the other reality can occur through the interpretation of signs (e.g., a rainbow signifies the Rainbow Serpent; a bull roarer's sound signals nearby spirit powers). Communication with the other side of reality also can consist of extraordinary initiatives taken by the spirit powers. The contents of these communications can include instructions in new dances, songs, and rites, and new magical powers being conferred on men who are severed from ordinary experience—in a dream, "trance," or some other nonordinary state of consciousness.

The Aborigines have a distinctive time consciousness that, of course, derives from everyday life but is also part and parcel of their belief system, which gives *meaning* to the routines of everyday life. All Aboriginal religious activities can be considered different but homologous means of reestablishing contact with and immersing oneself in the sacred time of the Dreaming. Every religious act, Eliade notes, "is only the repetition of an event that took place in the beginning of time" (1973: 84). There is in the Dreaming not a creation out of nothing but rather a conceptualization of space and time. The earth and life were already in existence when the extraordinarily powerful beings of this epoch began their work. During the Dreaming, enduring shapes were formed on the land, enduring connections were made between powers and places, and repetitive cycles were established.

Among the Pintupi, the life pattern of Aborigines living in dispersed, local groups makes Society a precarious achievement that is sustained through a projection of symbolic space beyond the local and the immediate (Myers 1986: 48). The structure of the Dreaming results from the way society reproduces itself in space and time. The distinction between the Dreaming and everything else is basic to the Pintupi view of reality. Both the country—the landscape and its significant features and objects—and the people are thought of as coming "from the Dreaming" (*tjukurrtjanu*), which is the very ground of being. The "Dreaming," to the Pintupi, refers both to the sacred stories and to the creative epoch of which the stories are a sacred history. There is an important semantic opposition that reveals a consistent ontology—an opposition that contrasts events associated with the Dreaming (*tjukurrpa*) with events or stories about that which is visible to the senses, to all that is called *yuti*. What is at issue is not whether an event is actually perceived but whether or not it is *in principle* witnessable. The relationship between these two levels of reality is not one of simple opposition; it is rather the case that the Dreaming is the ground or foundation of the visible, present-day world. The phrase "from the Dreaming, it becomes real" (*tjukurrtjanu mularrarringau*) thus "represents a passage between two planes of being" (Myers 1986: 49).

The conceptual bridge between the Aborigine and the world of the Dreaming is the totem. In Aboriginal totemism, people, natural species, and a vast range of natural phenomena come to be included, through processes of contagion and association, in one system of classification, as members of a moral, social, and ritual order. The spirit beings can be embodied in totems, which can range from winds, to plants, to animals, to bodily processes. For the tribal-living Aborigine, the totem is a sacred link to the Dreaming. One's totem is determined not by time but by location on the landscape. Thus if a woman conceives near the place of the sea eagle, the baby's totem will *be* the sea eagle, and in a sense the child will have a double existence, as a human and as an eagle. At the moment of conception, the human is the embryo of an ancestral being, a spirit-child that will develop into a divine human (Cowan 1992: 105–106).

In the process of several initiations, the member of Aboriginal society gradually advances to near-ancestor status as secret sacred knowledge is acquired. This advancement also depends on the negative cult, seen as a system of abstentions, and on involvement in a pattern of life. Among the Native American tribes, this patterning of life involved a severe asceticism that could entail the use of psychotropic plants, sensory and physical deprivation, and the ability to withstand pain (Durkheim [1912] 1965: 352; Elias 1993: 157–61). The Aborigines are milder in their asceticism, as what is involved are interdictions, constraints, and renunciation at initiation—ranging from fasts, retreats, unctions, and lustrations at the mild end of a continuum to tooth knocking, scalp biting, wounding with sticks red with fire, circumcision and subincision, lying on a bed of leaves covered with live coals, and fingernail tearing at the high end (Howitt 1904: 569, 604). However, initiates in this process are comforted, given whispered words of encouragement, held, stroked, and cosseted, and their fears are assiduously alleviated (Sansom 1980: 169). Through suffering, and scarifying of the body, there comes into being "the sign that certain of the bonds attaching the candidate to his profane environment are broken" (Sansom 1980: 355). Such severe tests—which are bravely met with disdain and which thereby demonstrate the courage and worthiness of the neophyte—are now less frequently practiced, largely because initiation is, for uninitiated Aboriginal boys and men, now apt to be optional: the prospect of excruciating pain is, as one might imagine, a disincentive to undergoing initiation.

Apart from the aforementioned, negation of the profane also requires a withdrawal from society, which means not looking at women and uninitiated persons but living in the bush under the supervision of knowledgeable old men, the elders. Altogether this systematic asceticism is the result of the hypertrophy of a negative cult. Participation in this negative cult exercises a positive action of great importance to the religious and ethical development of the individual. As Durkheim explained, because of the barrier that separates the sacred from the profane, it is possible to enter into intimate relations with the realm of the sacred only after ridding the self of all that is profane. Durkheim inferred, "He cannot lead a religious life of even a slight intensity unless he first, and in the context of formal ritual, commences by withdrawing more or less completely from the temporal life" ([1912] 1965: 347). There is thus *a temporal duality* separating the profane, temporal reality and the sacred reality of the Dreaming. The person who travels this path approaches the sacred by the very act of leaving the profane. "He has," Durkheim added, "purified and sanctified himself by the very act of detaching himself from the base and trivial matters that debased his nature" (*ibid.*). This process of sacralization is a lifelong practice, but it begins with initiation. The candidate for initiation passes long months interspersed from time to time with rites in which he must participate. He emerges from this process a member of the cult, so changed, according to Durkheim ([1912] 1965: 350), that he has undergone a virtual metamorphosis, and he is in some societies said to have been killed and carried away by the god of the initiation, which is, at least in the southeastern tribes studied by Howitt (1904: 67), apt to be named "Bunjil," "Baiame," or "Daramulun." The candidate either undergoes a second birth or, in some tribes, turns into quite another individual who takes the place of the candidate for initiation/manhood.

This metamorphosis is both social and spiritual: socially, there is the change from boy to man; spiritually, from participation in the profane, everyday reality to participation in the sacred secret, extraordinary reality. In the Aborigines' religion, and perhaps in most religions, introduction of the neophyte into the second reality, the circle of sacred things, requires that the person be separated from the profane world.

Penetration of the Present by the Past

The seven criteria being developed, as a definition of patterned-cyclical time consciousness, are interdependent and interpenetrating, understandable not as separate features but as an integrated whole or gestalt. Thus in turning to our next criterion, a fusion of the present and the past, we find the topic inseparable from the first criterion, and so it will go with the rest. Aboriginal religion is crystallized in myth. The tales are a statement of what was ordained at the beginning of the world. The chants and seemingly endless verses of the song cycles and "sing-song" dances during the ceremonies and *corroborees* (social dances) of the Aborigines "re-enact the dramatic essence of their tribal history—that vivid circular band that links the past with the present" (Gale 1980: 5). Stanner defines the Dreaming as "a kind of narrative of things that once happened; a kind of charter of things that still happen; and a kind of *logos* or principle of order" (1979: 24). And Lévi-Strauss (1966: 226–44), drawing on Strehlow (1947), sees that the everyday activities of traditional Aborigines are both reenactments of the *noumenal*, mythic period of the Dreaming and at the same time are contemporaneous with these same ancestral beings insofar as the ancestors are believed to be still engaged, invisibly, at sacred places and at sacred times, with ordinary, *phenomenal* life being seen as but a pale reflection of the more real world of the ancestors. Paradoxically, mythic history is "both disjointed from and conjoined to the present. It is disjointed from it because the original ancestors were of a nature different from contemporary men.... It is conjoined... because nothing has been going on since the appearance of the ancestors, except events whose recurrence periodically effaces their particularity" (Lévi-Strauss 1966: 236). This twofold contraction, Lévi-Strauss contends, is resolved by the joint effects of three kinds of ritual performances: (1) historical rites, which recreate the past, making it present (Past → Present); (2) death rituals, which recreate the present, making it integral with the past (Present → Past); and (3) rites of control, which adjust the abundance of totemic species in the present to the fixed scheme of relations between humans and totemic species that was established in the Abiding Law of the Dreamtime (Present = Past). It is not time that is destroyed, as Lévi-Strauss maintains, but the *effects* of time (see Gell 1996: 26–27). Lévi-Strauss, however, is quite right to qualify his analysis with a statement that the reconciliation of the "timeless" order of things that are *sub specie aeterniatis* and the disorderly nature of the everyday world are aimed at, but never fully achieved, by ritual means (Gell 1996: 27).

Several scholars have advanced the strong claim of a "fusion" of the past and present in the Aborigines' time consciousness. Basically this means a determination of the present by the past, which is brought about by the actions of people who "follow the Dreaming." As early as 1938, Elkin characterized Aboriginal values as collectivistic, as placing great value on the extended kin group, living in harmony with nature, and having no conception of time in the European sense—rather, living only for the present and remembering the past as an integral part of the present. Referring to the Dream Time, when the cult heroes and ancestors performed their deeds, Elkin suggested that this past is not "before" but rather "within" the present. That past "is present, here and now.... It is manifested in the present through the initiated (especially in sacred ceremonies)" ([1938] 1979: 232–34). Elkin argued that each cult group is the custodian of a set of particular episodes of the Dreaming and the associated rites. His hypothesized fusion of the past and present is focused on traditional Aboriginal rituals that reenact significant events of the Dreaming, as he writes: "Men in ritual may by means of... chanting, take the form of, and appear as, a natural species; and by doing so they become a medium of the dreaming. We who look and who hear the chants, are seeing a reenactment of a past heroic event; but to the Aborigines it is a "real presence"; it *is*—not *was* Dreaming" (Elkin 1969: 93). Elkin sees that the routes followed by Dreaming Heroes, and the features of the landscape marking their paths, are of the past. Similarly, Pam Harris explains: "The past is not gone. It is unseen, but becomes manifest in the present" (1984: 2).

Eliade (1959: 208) described the time of the sacred as not eternally bounded but rather "a primordial mythical time made present." He saw a common core in the Australian Aboriginal religions in their use of present-day ritual activities to establish "contact" with the Dream Time, which he located at "the beginning of time" (1973: 84). Every religious act, he maintained, is the repetition of an event that took place in the beginning of time. This psychic unity of past and present is realized in ceremony and ritual. At the moment of performance of a ritual, the persona of the living and the ancestors is not distinguished. Walbiri dancers are *both* "themselves" and ancestors while dancing, just as dreamers of ancestors are, for the duration of that dream, simultaneously ancestors and "themselves" (Dubinskas and Traweek 1984: 24). Munn ([1973] 1986: 33) suggests that designs and dancers exist as a representation of a distal other—the *djugurba* realm. But as Dubinskas and Traweek correctly point out, *djugurba* is not only the ancestral past but is in addition a socially acknowledged mode of experience in the immediate present. They see in the fusion of these two temporal realms an assimilation of time to space:

> It is the western cultural propensity to isolate "time" as an absolute, one-dimensional vector. *Djugurba*, however, is also a spatial location on the other (in)side of the earth's surface; it is the realm of the ancestors—those whom the dead become as they recede from living memory into a "past"; it is the source of energy for the growth of the contemporary world; and it is the narration or creation of that energizing as a social act. (1984: 24)

This idea of the Dreaming as a textual account is supported by Munn's ([1973] 1986: 112–17) study of Walbiri iconography as practiced by women in the casual, informal setting of the camp. She found that it is not enough, and not exactly accurate, to say that these stories tell about, or reenact the Dreaming; they are, she insists, rather *constitutive of* the Dreaming. The term *djugurba*, which means the ancestral inhabitants of the country and the times in which they traveled around creating the world in which the present-day Walbiri exist, also means "dream" and "story." *Djugurba*, the ancestral period, is outside of the memory of living persons and corresponds to what Schutz has called the pure "world of predecessors" (1967: 208). The term *yidjaru*, in contrast, refers to the world of the living, the ongoing present (Munn [1973] 1986: 23–24). The "Dreaming" here has the meaning of a story told on the sand. Such a story is a multimedia production: it exists as a visible icon, consisting of a sequence of drawings made, erased, and then redrawn in the sand; these drawings are produced with stylized hand gestures and narrated with an intonation that resembles singing. The Dreaming, in this context, takes place right on the ground, amidst the daily activities of the women's camp. Both the story and the marks on the ground are *djugurba*, which means, "Look, a story." The gestures of clearing a space to draw, and the erasures of drawings, which recur periodically as the story is told, structure the narrative and mark the temporality of the story (Masuzawa 1993: 170–71).

On the construction of time and space in the sand-story narration, Munn explains that as the story is recounted, successive graphic elements appear on the sand, bound directly to the flow of narrative action. The sequence of elements in the storytelling process gives a temporal order to the narrative, and the arrangement of elements on the sand reflects the spatial ordering of the story, this assemblage constituting a graphic scene in which "division between scenes is marked by erasure, and a graphic story develops through the continuous cycling of scenes in the manner of a movie" (Munn [1973] 1986: 69). Here, indeed, we see the Aborigines' sacred, inner reality manifesting itself in the everyday world of tribal women. The ancestral beings referred to in these stories are, according to Munn, "anonymous persons ... who lived in those times, rather than identifiable ancestors" ([1973] 1986: 77). When asked where the story took place, the storyteller would answer "*Djugurba*," in this context, meaning "no place." Here we have no instantaneous "fusion" of the Dream Time and the present; it is rather the case that *djugurba* begins to spill over into the "here," "now," and "everyday." Munn explains that the narration, taking place amidst everyday routines of camp life, is itself part of everyday reality.

Deborah Rose, in her study of the Yarralin of the Victoria River Valley in the Northern Territory, found a clear dichotomy between Dreaming life and ordinary life, suggesting that "One way of expressing this distinction is temporal: Dreaming precedes ordinary. Ordinarily life belongs to the present in which we now live. It is characterized by temporal sequence and is marked by beginnings and endings" (1992: 204). A fundamental distinction between these two kinds of time has to do with the distinction between permanence and impermanence. That which is Dreaming does not die and does not get washed away, rather existing forever. And as

Myers writes, "The 'mythical' transformations of the Dream Time ancestors into the land and its features have the form of a 'permanent atemporal identification' of ancestors with the land. These acts still influence the present" (1972: 9). Rose quotes one of her informants who refers to the transformational continuity and endurance of the Dreaming:

> That thing [the Dreaming] never get away. Blackfellow there, he's still there. Nother man taking over. Nother bloke taking over for him. ... You see, my thing, you never take him out. He's there for years and years. People been die, and still him there. And nother bloke takes over. Nother man die, he's still there in the ground. From beginning. When him from beginning. (1992: 204)

Rose found that most Yarralin trace their genealogies back to their grandparents and great-grandparents, who are generally believed to come directly from the Dreaming. Thus for this Aboriginal society, the Dreaming happened about one century ago. She cites Old Tim, who believes that his mother was of the Dreaming, saying that she "been come out from the ground, underneath from the ground inside, been come out" (1992: 204–205). While the dreaming is placed about 100 years ago, this distance does not increase over calendrical time, because the reference point is not any fixed time of the Dreaming but rather must be understood as the present, the "now." The Dreaming is not lost to the past but rather coexists with, moves together with, the present.

HETEROGENEITY, DISCONTINUITY, IRREGULARITY

Durkheim's ([1912] 1965: 114–15) methodology for the study of totemism sought not exterior resemblances but rather an understanding of totemism in its own actual and real contexts, with his focus on the ethnographic studies of the Aranda and other tribes of central Australia. The sacred objects of the clan (e.g., the Aranda's *tjuringa*) are rocks or pieces of wood, made sacred and powerful by having the emblem of the clan engraved upon them. The *tjuringa* contains the image of the soul of an ancestor, and it is the presence of this soul that confers on the *tjuringa* its sacred powers (C. Strehlow 1907). The emblem on the *tjuringa* is more sacred than the *tjuringa* itself (Durkheim [1912] 1965: 156).

An Aboriginal man bears the name of his totem, this name being an essential part of his very being. A member of the kangaroo clan calls himself a kangaroo. Durkheim inferred that "each individual has a double

FIGURE 3.1 Two Aboriginal men telling a story about the sacred history of a stone *tjuringa*, Haasts Bluff, Central Australia.

nature: two beings coexist in him, a man and an animal" ([1912] 1965: 157). Insofar as the two realities of *homo duplex* are separate, we have a discontinuity. This discontinuity can be seen in Aboriginal beliefs about the origins of humanity. Beliefs differ from tribe to tribe but generally refer to a violent disjunction as dreaming creatures were changed into humans. For example, in the initial conditions of the world in Aranda cosmology there existed on the earth's surface the *inapatua*, which were chained together masses of protoplasmic humanity, a semiembryonic mass of half-developed infants. These bodies existed without sensory experience and were fused together without individual identities. Then, as the story goes, a Lizard Ancestor cut them into individual people, shaped their organs with a knife, subincised and introcised them, and gave each a personal *tjuringa*. The *inapatua* had undergone their first initiation. Durkheim noted that when a sacred being subdivides itself, it remains whole and equal to itself, so the part is equal to the whole. This means that sacredness is not dependent upon physical law. Rather, its virtues come from sentiments and mental processes that, as Douglas puts it, "float in the mind, unattached, and are always likely to shift, or to merge into other contexts at the risk of losing their essential character" (1975: 49) under threat from the profane and from the sacred but evil powers of the Dreaming, a threat that is met through rite and interdict. Particular sacred things, then, are but individualized forms of natural, impersonal forces. We will see in chapter 9 that these disconnections and discontinuities create the opportunity for synthesis, a key capability of the gestalt synthetic mode of information processing of which patterned-cyclical time consciousness is an aspect.

Event Orientation: The Happening

> In the popular Western view, time still, so to speak, ticks on even if nothing occurs; its emancipation from events is ensured by its own subjugation of an ongoing numbered measure. But in Aboriginal thought, there is nothing beyond events themselves. This is entirely apparent in their cosmology, which lacks any reference to ultimate pre-event origins. For Aborigines, there is nothing more fundamental than the statement: events occur.
> —Tony Swain, *A Place for Strangers*

Ancient people, at least in the Mediterranean, did not separate time from its contents. As Ariotti explains, "For them time was [a] qualitative, phenomenological, concrete interval. ... Time was not a neutral and abstract frame of reference. Time was its own content. Events were not *in* time, they *were* time" (1975: 70). This is the singularization of time. The association of time with singular events also characterizes the ancient people of Australia, and no doubt many other oral, indigenous cultures, peasant cultures, and, perhaps, cultures of the ancient world in general.

In modern societies, Western and non-Western, detailed, written records of the time at which events take place are recorded, and historical markers, cairns, and obelisks are constructed to commemorate significant events. Western, Gregorian, calendrical time is based on an event, the birth of Jesus Christ. Moslem calendars are dated from 622 *Anno Domini*, the year in which the prophet, Mohammed, fled from Mecca. As Pam Harris (1984: 17) points out, the passage of time is everywhere marked by events.

Bull (1968: 3) avers that every event happens in time, every event uses time, and it is as we observe events taking place that we notice that time is passing.

Aboriginal religion is based on events. The "Abiding Law" (Swain 1993) is no less than a life plan derived collectively from Dreaming Events. The still-abiding events of the Dreaming are coterminous with life events, particularly those events carried out in formal rituals. For an Aborigine to say "That mountain is my Dreaming" means that this particular mountain holds secret knowledge and moral truth (Gondarra 1988: 6). Swain argues that precontact Aboriginal ontology depended upon a class of events—the Abiding Events—collectively constituting the Abiding Law, the norms and rules for living in an Aboriginal community. In situations where Westerners would locate events using ages and dates, Aboriginal people relate events to other events that happened at about the same time. Rose sees in Aboriginal temporality a linking of events to specific places, as she writes: "Temporal coordinates can be fixed through social events: when so and so was made into a young man; when Big Sunday (an important ceremony) was held at such and such a place" (1992: 203).

Among the Yolngu, Rudder (1993; also see Keen 1994: 42) found that an event is located by one of four methods. First, it can be located by a spatial metaphor (e.g., *barrku birr!*, "far off"). Second, it can be located by terms referring to an imprecise far distance in time (e.g., as *baman'*, "long ago," or as *baman'birr*, "very long ago." The third practice is to refer to singular historical or religious events, where certain conditions obtained (e.g., *mitjinmirriy*, "the times when towns were staffed by missionaries"; *womirriy*, "during World War II"; *wanjarr*, "creator ancestor time"). And fourth, events can be described with reference to stages in the life of an individual person. For example, a man might speak of first seeing certain sacred objects at his own initiation ceremony. Here his focus of attention is on the location of the event in this stage of life, with distance from the present only of peripheral interest. Rudder (1993: 272) lists twelve such life-course events, describing for the self, father, and grandfather the events of being born, being circumcised, the first son being born, and participating in important ceremonies. The temporal term *baman'* stands in contrast both to the present (*dhiyangu-bala*, "now") and to the past of living memory of remembered ancestors. Genealogical distances can also be invoked: "My father did not know the *wanjarr*, his father did not know, his father did not know; long ago" (Keen 1994: 42). Warner's ([1937] 1958: 566, 568) study of this cultural group in the 1920s found informants using much the same expressions. He took the term *baman'* as the name of an ancestral period in the same way that more recent scholars have interpreted the term *wangarr* as a time category. Warner defined *wangarr* (his "*wongar*") as a general name for the totemic spirits, while "in the time of *Wongar*" referred to the mythological period "*baman.*" Morphy (1984: 17) and Williams (1986: 28) both see *wangarr* as referring to a distant time in the past and to the spirit forms that existed during this period. Keen, however, sees this definition as misleading, because the Yolngu language has no abstract category "time" and "so did not play this language-game of parallel domains of time" (1994: 43).

In contrast to Westerners, Australian Aborigines traditionally do not keep track of a child's age beyond a few years. Now, however, knowledge of age and of one's birthday is widely known by Aborigines, as a result of their experiences with missionaries, health workers, government officials (who insist on keeping records of births, deaths, marriages, and divorces), and schools and other formal institutions of the modern world. Yet even today, Aborigines prefer to approximate age by reference to a person's size and maturity at the time of significant events. Pam Harris (1984: 15) provides an example of an assistant teacher at Lajamanu who described places where she had lived as a child, indicating the length of time at each place by showing how high she was at each part of the journeying. She said: "When we left Gordon Downs I was still in my coolamon and then when we left Montejinni I was this big, but I still got carried some of the time." Thus we see that while in the wider Australian society there is an emphasis on dates and ages, in Aboriginal society, events used in reference to a person's life stages are the focus of attention, with distance from the present a peripheral aspect of this location and with the continuum aspect of time irrelevant. An Aboriginal student at the Batchelor College teacher training course summed up a discussion of time and the marking of its passage by saying, "Europeans record, Aborigines remember" (P. Harris 1984: 18).

The Aborigine's "Dreaming" is not based on ordinary, clock time, but this Western notion of time can occur in the content of ceremony and ritual. Among the Yarralin, song verses and accompanying dances often are punctuated with a running commentary of verbal and visual jokes, many of which draw on the Western concept of time and address the question "What time is it?" Rose (1992; also see Thomson 1935) provides an example:

> Responses to the question take the form of elaborate jokes which mock European concerns with time. Various people call out for a wrist watch; they query whether it has been wound or whether the batteries are fresh. More people call for a torch so that the watch can be seen. The person who calls out the time will be criticized for not being able to tell time, and other people look at the same watch and call out different times,

while others offer time at random, or make up times (28 o'clock). Joking goes back and forth until a time is agreed upon. Perhaps an hour later the same process is repeated, and the agreed time may be half an hour earlier than the first time, or three hours later. (216)

The Aborigine's "Dreaming" does not refer to any kind of "time" that can be compared to "normal time." In order to understand the meaning of the Dreaming, it is not enough to ponder the proper translation of the Aranda term *aljiringa*. Instead, it is necessary to realize that, in Swain's terms, "the Aborigines constructed their world in terms of *rhythmed events*," and that the Dreaming in particular is a "class of events" (1993: 22). Swain follows C. Strehlow's rendering of the Aranda *aljiringa* as "eternal events" and suggests that the Dreaming consists of "Abiding Events" which has the verbal significance of "residing in a place." Swain proposes that the precontact Aboriginal world consisted of the interplay of two kinds of events: rhythmic and Abiding.

The Aborigines have traditionally believed that before the Dreaming, before *the primordial events* had occurred, the pristine landscape was but a formless void. These world-creating events still occur whenever there is Aboriginal participation in them through ceremony and ritual. The Dreaming event had been final and complete, so orderly in its sacred perfection that there was no need for the land itself to change or develop. Aborigines use the land as a mnemonic device to help them recall events from the Dreaming.

The records Aborigines keep of events that happened in the long-distant Dreaming are not like Westerners' written records and dates. Instead, each and every tribe has Dream Time stories, ceremonies, sacred sites, and well-known landmarks that remind them of their own origins and of the activities of their totemic Ancestors and Primordial Heroes. Detailed accounts of these events are kept in the collective memory of the people and passed from generation to generation. "It is the events, and their significance for the life of the people now, that are important, not the precise time at which they happened" (P. Harris 1984: 20–21).

The centrality of events to the time consciousness of Aborigines can be seen clearly in Rudder's (1993) analysis of Yolngu notions of time, which are concerned with: the location of events in relation to other events in time and space; the relationships between events; the power involved in events; and duration, as an assessment of the effort or power involved in events. It is the continually repeated events of the natural world, with the day and season given the most significance, along with the movements of the moon, stars, tides, and winds. Consider, for example, the day, which is segmented into eleven periods, from sunrise, "the sun throwing its hands out" (*gon-djalkthunminyaray*), to "the sun shining on the side of the head" (*läyyu-wail*), to "the sun going in" (*walu-gäarrinya*). A similar collection of terms describes the night, such as "second morning star rising" (*dhudi-djaw'yunaray*). This classification is focused not on the positioning of events in a series but on events themselves. Rudder further suggests, "Events of a mythological nature, irrespective of their antiquity, almost cease to be events in time at all. By becoming part of the inside, they are then part of the unchanging eternal which is past, present, and future" (1993: 278).

Sansom, in his study of a "mob"[1] of Aborigines living in the fringes of Darwin, found them to "recognize happenings as the basic unit of social action, and one proof of this is that their "happenings" are labeled entities" (1980: 3). As an example of the Aboriginal English of the camp, the statement "Somebody bin take beating" will evoke in the Aborigines "a total procedure, a staged performance with a typical beginning, painful middle, and standard ending" (Sansom 1980: 3). A happening is the form that people bring to the flow of social action. Participation in this action is structured by culturally provided rules governing proper performance. Happenings are more than instances of behavior, for they are events that are true to type and can therefore be analyzed to reveal significance transcending the moment of an event's enactment. To miss out on events that involve significant trouble for the collectivity is to be subject to reproach. To miss significant happenings, such as a mortuary ceremony ("dead body business"), can be a "mistake" or a dangerous "wrong thing," involving a social debt that can demand redemption and that can become an unfortunate turning point in a societal member's life, perhaps resulting in his or her being ostracized by the group, encouraged to "go another way" (Sansom 1980: 145).

Chapter 4

Patterned-Cyclical Time Consciousness, Continued

In this chapter we consider the three remaining aspects of patterned-cyclical time consciousness: time as cyclical and based on patterns and oscillations, qualitative time reckoning, and the experience of long duration.

Time As Cyclical, Patterned, and Oscillatory

To live in harmony with the world is to be attuned to the rhythms, cycles, and patterns of the natural world. Evans-Pritchard (1940), in his study of the Nuer, referred to this kind of "cyclical" time as "oecological." The notion of patterned-cyclical time being developed here, in contrast, is oecological but is *also* based on the rhythms and cycles of biological development, social life, and cultural patterns, all of which, as will be shown in chapter 10, are mentally grasped as a whole or gestalt. For Aborigines, the life of the individual person cycles from entry into the womb, through birth, infancy, childhood, maturity, old age and death, and, according to the doctrine of many but not all Aboriginal tribes, to reincarnation in a potentially endless cycle. If people properly "look after" country, country will form an essential part of human survival.

One way in which Aborigines experience the landscape and environment is through carrying out Dreaming Journeys. These Journeys can be carried out as a practical, social activity involving the family or band. The Dreaming Journey is a going outward into the physical and social world but is a cycle insofar as the outward journey, regardless of its duration, culminates in a return to a point of origin. The tribal member in such a journey seeks the experience of spiritual renewal through contact with the mythologized landscape. The seasonal cycle is a practical reason for undertaking such a journey, which is articulated with food gathering and hunting activities. During these journeys, contacts are made and renewed with members of other tribes and with friends and relatives. These journeys are also, on occasion, carried out alone, as a "walkabout," in an effort to deepen understanding of one's own sacred nature. In this process the metaphysical landscape is transformed into a hagiographic history of the people's origin. The primordial events that took place during the "Great Time" or "Dream Time" of creator beings, even before their ancestors were born, mark the landscape and continue to occur even as the individual seeker makes his or her Dreaming Journey. When the landscape is fully humanized, it becomes appropriate to speak of the "totemic landscape," in which the Aborigine is linked to the topography of the landscape by the spirituality encoded in the mythology of the Dreaming (T. G. H. Strehlow 1970; Myers 1986). The topographic landmarks, or sacred sites, that the Aborigine encounters on a Dreaming Journey are noetic points contingent to a mythologized landscape that is seen as having been constructed by the Primordial Heroes.

Strehlow writes that for Aborigines, "there is no division between Time and Eternity. Since *every* person carried within himself or herself, through reincarnation, an immortal spark of life derived from the original supernatural personage, men and the totemic ancestors were believed to be interlinked inseparably" (1970: 132). Thus the visible totemic landscape is an integral part of the reality of eternity. The Aborigines believe that they must live, as best as they are able, in harmony with nature and the land, attuned to the rhythms and patterns of the cosmos. Every species of plant and animal is seen as having its right to exist, not necessarily for human use but for itself. The organization of life is a network of interdependence. Writing of the Yarrilin of the Victoria River Valley (Northern Territory), Rose writes: "Yarrilin people believe that if they take care of country (burn it, perform increase rites, sing the songs, visit and use it), country will take care of them. This is a cyclical process in which the knowledge and care which humans put into the system ... form an essential part of human survival in the system" (1992: 101). In Aboriginal theology, the earth as much as mankind was required to participate in ritual acknowledgment of its sacred existence in order to "live" and "bear fruit." Tribal-living Aborigines are deeply concerned that the ceremonial cycle each year should be completed correctly and with due reverence. Were this not to happen, the earth would "harden" and fail to enact its fructive potentiality.

The relationship between the Aboriginal person and country must be built up over time and with the assumption of responsibility. A person is born with rights, Rose (1992: 108) notes, but each person must choose to further develop her or his own relationship with country. Damage to country can bring death or injury to people, and damage to people can damage country. Rose provides an example of the latter in the form of a story told to her by an informant: "When Allen Young's father died, a Walujapi (black-headed python) Dreaming tree broke in half, and the water it contained all ran out" (1992: 108).

At the time of birth, the male Aboriginal child is unaware of his previous existence. Later the elders will initiate this child and reintroduce him into the ancient ceremonies that he himself had instituted in his previous existence. Through initiation, the novice discovers that he has already been here, in the beginning, during the Dreaming. In learning the deeds of his mythical Ancestor, he learns about his own glorious preexistence. This Aboriginal doctrine of anamnesis has a Platonic structure, as for Plato learning is recollection, and to know is to remember (cf. *Meno* 81). When the soul is reincarnated, it drinks of the spring of Lethe and forgets the pure and perfect knowledge it had obtained. In order to remember his preexistence, the Aborigine must undergo initiation and must also withdraw into the inner world of contemplation in an effort to rediscover and repossess its knowledge and experiences of the Dreaming (Eliade 1963: 124). For the Aranda, the knowledge in question is not philosophical but mythical and historical, as the initiate learned what he himself did *illo tempore*. But for both Plato and the Aranda, the true anamnesis is the result of spiritual activity (Eliade 1973: 59). Eliade saw the paradigm of the Dreamtime as a key not only to understanding the Australian cases but also for all other "primitive" religions. In his view, the Dreamtime is but an Australian version of the mythic beginning, as it explains how the various kinds of primordial beings appeared at the very beginning, what they did, and why and what happened to them.

John Morton (1987), building on the work of Carl Strehlow (1907) and his son, T. G. H. Strehlow (1933, 1947, 1956, 1964, 1971), on Munn's (1970) studies of the Walbiri and the Pitjantjatjara, and on some helpful commentary by Myers (1986) on the Pintupi that we have already considered, provides an excellent example of the cyclical nature of life and death among the Aranda. Here it is sufficient to display his two models of "transcendental reproduction."

Aranda ancestral figures are seen as containers of innumerable "cells," all of which are "spirit children" (*ngantja*), each a potential duplication of the ancestor himself or herself. A man's soul, his *atua ngantja*, always first manifests itself by entering a woman from a *tjurunga* "hidden" on the land by the ancestor (C. Strehlow 1908: 1–56; also see Spencer and Gillen 1925: 75–77). A tribal-living Aboriginal hunter might spear a fish and then realize that this creature is inwardly a preexisting child-spirit who will follow him to enter his wife's womb. Evidence that this has happened is apt to be found in an ordinary dream, usually but not necessarily by the husband, in which the spirit-child has entered his wife and will be born as the husband's child (Kaberry 1939: 278). The conception totem refers to an object through which a spirit-child reaches a woman's womb. Kaberry contends that spirit-children beliefs explain conception, provide for the child a totem—perhaps but not necessarily a water hole in the father's territory—and also provide a spiritual link to the Dream Time.

Morton's schematization of these relations is shown in Figure 4.1 (panel A) in terms of an opposition between individuation and fragmentation and maps the opposition onto a continuous circle—meant to suggest that the two movements (unity to multiplicity, multiplicity to unity) are complementary. This opposition is homologous to another, as follows:

> one : many :: old : young (ancestors are identifiable with senior men, while spirit children, or the *inapatua*, are, of course, infants). … [T]hese contrasts make Diagram [4.1A] nothing less than a representation of the skeletal structure of the Dreaming itself, implying a form of transcendence existing within a context of spatiotemporal relations (Aranda society). This transcendence, I suggest, is the equivalent of Munn's *a priori* grounds upon which the possibilities of … order are built. (Morton 1987: 113)

Morton's second diagram (Figure 4.1, panel B) summarizes the religious life cycle of an Aranda man. Point 1 is where life begins as a spirit-child. When the spirit-child (moving counter-clockwise on the diagram) enters the mother's womb, it leaves behind its *tjurunga*. When the child is born and rises above the earth's surface, it cries for its lost *tjurunga*, which is then brought out and placed in his cradle (Strehlow 1908: 80). The baby is in this period a *ratapa*, an age term covering in utero existence and early life outside of the womb. At about one year after birth, the baby crawls away from the mother, is given a name (Strehlow 1913: 3), and is recognized as a social being. At this point, the *tjurunga* is returned to its hiding place. The movement from point 1 to point 2 represents to the Aranda a loss of their spiritual roots; during childhood, the period from points 2 to 3, the child is close to his mother but is ignorant of the esoteric aspects of

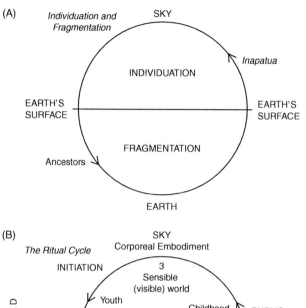

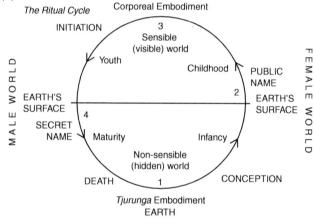

FIGURE 4.1   (A) Opposition between individuation and fragmentation mapped onto a continuous circle, suggesting that the movements unity-to-multiplicity and multiplicity-to-unity are complementary; (B) The religious life cycle of an Aranda man, from point 1 (life as a spirit-child) to point 4 (death).
Source: Morton 1987, diagrams 1–2: 113–14, Reprinted with permission of John Morton.

the Dreaming known to the adult men. This is a period of unheeding "ignorance" (see Myers 1986: 107–108 on the Pintupi).

By adolescence (at point 3), the boy begins his initiation process, during which he rediscovers his spiritual past. The journey from points 3 and 4 represents the process of systematic initiation into the male cult. This, as Morton explains, "represents first of all the 'filling up' (introjection) of a novice during the repressive, initiatory ordeals, and second the 'discharging' (projection) of ritual obligations in a movement which ends in death" (1987: 115). This, however, is no ordinary death. Instead, as Munn (1970) argues, there would appear to be a strong belief that an incarnated, living spirit-being is metamorphosed into a feature of the countryside. The word referring to these spirit-beings, *burga-ri* (or *bugari*), does not mean "dying." It is rather the case that the spirit becomes the country, or a particular piece of the country. Generations of dead are thus successively transformed into country.

Returning to Figure 4.1B, the horizontal line separates the earth's surface and the underworld. Morton's focus is not on what we have called the "split" in Aboriginal time consciousness but on spatial correlates—between the "surface" level above and the "deep" level below. This is a model of sociotemporal order, so that neither place nor temporality can be dismissed out of hand, and both have descriptive value. Morton (1987: 115; also see Rudder 1993; Myers 1986) makes an important and a useful point, however, when he argues that the key idea involved is neither space nor time but *sensibility*. The earth's surface is the visible world of sense experience, whereas the underworld is hidden from the senses. Both "the distant past" and "the unplummable depths" are beyond what Myers calls the subject's field of sensory perception but capable of manifesting themselves in visions and dreams.

Natural and Social Cycles

John Harris (1979: 101) reports that Anindilyakwa Aborigines of Groote Eylandt distinguish two seasons in their monsoonal climate—the wet season (*yinungkwura*) and the dry season (*mamarika*). One year is seen as one wet season and one dry season. Even non-Aboriginal people in this region refer to the "last wet" or "next dry." The lunar cycle also is observed, as it is important for activities such as fishing. Further seasonal weather times are distinguished—rain time (*akilarrkalyukwa*), cloud time (*akilarrkaburarruwa*), time of the northwest wind, cold time, mist time, and hot time. An even more detailed cycle of seasons is distinguished in terms of available foods—from "yam time" to the "time of *dumungkawardarra* flowers," indicating the approach of the rainy season. One Aboriginal seasonal cycle is shown in Figure 4.2.

There also are terms for day (light, *menungwulida*) and night (black, *alyarrngwalyilya*) and eight terms for a more detailed daily cycle, from predawn (*buwawiya*) to midnight (*marngkirngkuwilyarra*). Instead of a

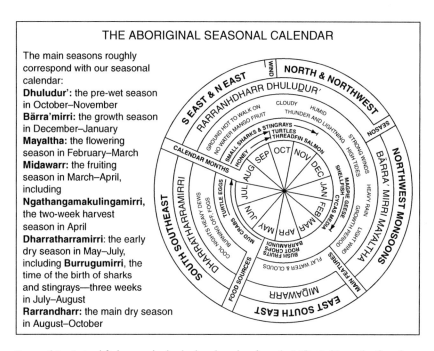

FIGURE 4.2 A simplified seasonal calendar based on data from the Milingimbi people of northeast Arnhem Land. The months of the year have been included for orientation purposes.
Source: Davis 1989: 2.

clock to fix an event in time, Groote Eylandt Aborigines might point to the sun's previous position and say, for example, *numiyamina manawura*, or, "the sun was like this." According to Whitrow (1989: 15), Aborigines might fix the time for a proposed action by placing a stone in the fork of a tree so the sun will strike it at the required "time." In some Aboriginal languages, such as the Anindilyakwa language of Groote Eylandt, either "sleeps" or "suns" can be used interchangeably. In the hunter-gatherer's life, there is no desire, for instance, to find new and better varieties of yams or edible seed; existence depends on the repetitive cycling of the seasons, of the growth of yams in their usual places, of the "increase" of kangaroos and other animals in their regular time and place, and so on. In Aboriginal languages, the same word "moon" is apt to be used for both the moon as an object in the sky and for talking about the span of time indicated by the moon. In the Kuuk-Thayorre language area of northern Queensland, "three months" would be described as *kapir yuur pinalam* (literally, "moon finger three").

Kinship and Ceremonial Cycles

There is an important and a necessary linkage between totemism and natural resources for tribal-living Aborigines. Aborigines see deserts and other regions as crisscrossed by the routes of totemic ancestors. These travel routes provide points, or centers, of social exchange and social contact between totemic clans and local groups. Many Aboriginal societies have historically enjoyed great ceremonial centers. Consider, for example, the Aranda. Survival of their cultural order depends on the ceremonial cycle of the year, which in turn is contingent upon the availability of water. The greatest sites of these ceremonial centers have been established where full-scale festivals could be held only in excellent seasons, with abundant water. T. G. H. Strehlow (1970: 95) points out that the three greatest Lower Southern Aranda sites were accessible only after exceptionally heavy rains had fallen over thousands of square miles in this area, filling the flood-flat depressions and large local swamps with water. Some ceremonial cycles last four to six months, during which time the irregularity of the water supply and of game and plant food can become problematic. To this end, each local group was responsible for holding increase ceremonies to bring about abundance of their own totem—whether rainfall, kangaroos, carpet snakes, or grass seeds, and any other useful plant totem. Each Aranda local group has the duty of staging, at irregular intervals, the complete ceremonial cycle associated with its own ceremonial center.

The cycle of life and death is most clearly seen among females, expressed widely among Aboriginal tribal groups through the four-section and eight-subsection systems (Barden 1973). Subsections are a kind of social classification unique to Northern Australia in which societal members are partitioned into four or eight named units known locally as "skins" (McConvell 1985). In subsections, a woman's daughters' daughters' daughters' daughter returns to her subsection in a four-generational incarnation cycle. In this system the emphasis is not on change and development but rather on sameness and repeated patterns of the known. Barden suggests that in the four-section system of desert-living Aboriginal tribes, all four sections are continuously filled, so classification is synchronic rather than diachronic. He poetically infers, "In the desert section there is not time at all; duration is overcome by the constant repetition of the pattern" (1973: 332–33). The author would take exception to any literal claim that time has no meaning in the desert. It should in this connection be pointed out that clocks and calendars today *do* have value even in the desert, although chronometry there is hardly of central concern. The advent of short-wave radio and satellite

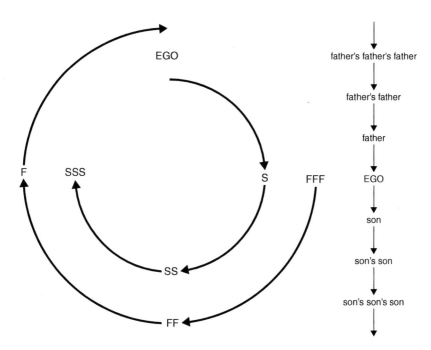

FIGURE 4.3 (A) A cyclical view of succeeding generations; (B) A Western, linear view of succeeding generations.
Source: Yallop 1982, figures 6.1–6.2: 147–48.

television throughout Australia—with Aboriginal programs scheduled at specific times on certain channels—has certainly sensitized these Aborigines to clock and calendrical time.

The cyclical nature of Aboriginal social organization can be found in kinship terminology. Aboriginal kinship terms do not emphasize actual blood relationships. The focus is rather on the overall structure of the tribe, clan, and extended family. In some tribes, as Yallop (1982: 146–52) reports, an Aborigine might apply the same term to both his father's father (FF) and his son's son's son (SSS). In some tribes, a member's son's son's son might even be "my father" (F), and his father's father's father (FFF) might be "my son." This suggests the view of successive generations that can be represented cyclically. "EGO" is the individual whose kin are named. Kin who are adjacent in the circle shown in Figure 4.3 are referred to by the same term:

son(s) = father's father's father (FFF);
son's son (SS) = father's father (FF);
son's son's son (SSS) = father (F).

Yallop further explains, "If EGO needs to refer to individuals equivalent to himself in this system (i.e., son's son's son's son or father's father's father's father), he will call them 'brother'" (1982: 147). This system is at odds with the Western system of linear descent, in which from the point of view of Ego there might be a potentially unlimited succession of children, grandchildren, great-grandchildren, great-great-grandchildren, and so on.

Qualitative Time Reckoning

Aboriginal thinking can be described as qualitative as opposed to quantitative. Rudder (1983: 6), using the *Oxford English Dictionary* (1971), defines quantitative evaluating as the assessment and measure of things in relation to one another in terms of size, magnitude, amount, duration, and extension. Qualitative evaluation, in contrast, is concerned with the attributes, properties, special features, and characteristics of things. Seagrim and Lendon, in a study of central-desert, Aranda Aborigines, found that they rely on qualitative cognitive evaluations, and that even material "exchanges that occur ... involve qualitative not quantitative equivalences" (1980: 201). Rudder makes the strong statement that the Yolngu do not use any form of quantification or measurement, and that all of their evaluations are qualitative. Thus it appears probable that a focus on qualitative evaluation characterizes Aboriginal people in general, which can be contrasted to the emphasis on quantitative evaluation found in modern, Western society, and far beyond. In this qualitative reasoning, "time" and "timing" are not considered particularly important, and punctuality is not insisted upon. Appointments are not attended promptly. "Aboriginal time," in jest, is referred to as being "one hour behind European time." Time for Aborigines is not valued for itself but as an opportunity to improve and pursue interpersonal relationships and to enhance a feeling of well-being (Duncan 1980: 53).

Swain writes that "the rhythms of Aboriginal life are totally unnumbered, as Aboriginal tradition uses no counting at all" (1993: 18–19). In the modern, Western world, there is an emphasis on age counted in years, but in Aboriginal society, age is less a property of individuals than it is groups of people belonging to ritual-status cohorts. As P. Harris puts it, "The Western tradition emphasizes *quantity and measurement*, whereas the Aboriginal tradition emphasizes *quality and comparison*" (1984: 17, emphasis in original). Even in the desert, where the ideas of dividing, measuring, and precisely quantifying time are not part of Aboriginal tradition, the intrusion of Euro-Australians into nearly every aspect of life means that they must, in the educational system and elsewhere, become acquainted with Western ways of measuring time (see P. Harris 1984: 17).

The qualitative nature of Aboriginal time reckoning can be seen in Rudder's (1993) analysis of Yolngu talk about events located in the past and the present. Using examples from two Yolngu dialects, Djambarrpuyne and Gupapuynu, Rudder shows that distance in temporal space is apt to be indicated not by reference to the measurement of time but rather to the location of the event under discussion. In cases where distance is measured, there is no fixing of events at any absolute time. Distance in time is rather measured as distance from the present. In these dialects, terms are used express distance in either geographical or temporal space (e.g., *galki* (close/soon), *marr galki* (fairly soon/fairly close), and *mirithirr galki* (very close/very soon).

In terms of seasons, the potential for locating events with precision is limited to the current year and the season with the same name in either the preceding or the following year. Other seasons in the preceding or following year can be located in time only imprecisely, as adjacent or not adjacent, and seasons two years removed in either the past or the present cannot be located in time with any precision. Rudder's conclusion is that regardless of the temporal location system that is used, events can be located accurately in time only in relation to the current repeating event, or the adjacent event. Beyond that, only imprecise location is possible. Thus traditional Aboriginal time reckoning is fully qualitative and not apt to involve the use of counting numbers.

THE EXPERIENCE OF LONG DURATION

> If we were to locate ourselves, hypothetically, in
> Dreaming, we would see a great sea of endurance,
> on the edges of which are the sands of ordinary time.
> —Deborah Rose, *Dingo Makes Us Human*

"In general," Chen writes, "time is never experienced as a whole" (1992: 40). As a progression of moments, as pure succession, the future has become the present and will soon be the past. This view of time denies the reality of future and past, which must inhere in the present for anything to exist. Chen (ibid.) concludes, "The fact that things exist through time demands a conception of permanence: duration." The problem is that once duration is eliminated from the succession-of-moments view of time, it can only be visualized and symbolized by extratemporal concepts—by a space-time continuum, or by eternity.

Without succession, duration is but an unchanging present, therefore, duration adequately describes only the flux of time. On the other hand, it is only within the background of duration that we can be aware of duration. "Duration then becomes not a static but the sustaining quality of time, and therefore the actual 'co-existence' of past, present, and future, a 'co-existence' which does not imply simultaneity because succession is included in their relationship" (Chen 1992: 152–53).

In his study of the Native American Hopi language, Whorf (1941) wrote that Hopi time reflects a way of thinking that is closely associated with subjective awareness of duration, of the ceaseless "latering" of events. To the Hopi, for whom time is not a motion but "… a getting later of everything that has ever been done, [so that] unvarying repetition is not wasted but accumulated. … [It] is as if the return of the day were felt as the return of the same person a little older but with all the impresses of yesterday, not as 'another day', i.e., like an entirely different person" (88). Such sequences of events are not positioned as they are in ordinary-linear time, on a smooth-flowing continuum in which everything proceeds at an equal pace but are rather accumulated and amalgamated. Whorf based his case on a study of the grammar of the Hopi language. It is possible to find this concept of value without endorsing Whorf's ambitious linguistic relativity theory and without rejecting the Chomskyian notion of transformational grammar.

The dichotomy in Aboriginal thought between the time of the Dreaming and ordinary time corresponds to the distinction between the enduring and the ephemeral. Durkheim ([1912] 1965) wrote that the passage of "time, by itself increases and reinforces the sacred character of things. A very ancient *tjuringa* inspires much more respect than a new one, and is supposed to have more virtues" (313). It is as though the veneration it has enjoyed over the generations has somehow accumulated. Dubinskas and Traweek (1984) define "duration" for the Walbiri Aborigines as the totality of "meanings coalesced in [the word] *djugurba*, the ancestral realm, and *yidjaru*, the ongoing present, as an interpenetrating symbolic complex" (29n). They specifically reject the use of the essentially linear, Western notion of "time" as a gloss for this meaning nexus, as it is deeply rooted and specifically elaborated in the Western cultural context to evoke the meanings suggested by the "text" of Walbiri culture (Dubinskas and Traweek 1984: 29).

The origins of the present, ordinary reality are to be found in the Dreaming but are characterized, in Rose's terms, "by changes which do not endure, by sequences which can be accurately described in temporal terms, and in the obliteration of the ephemeral" (1992: 205). The Dreaming, in contrast, "can be conceptualized as a great wave which follows along behind us obliterating the debris of our existence and illuminating, as a synchronous set of events, those things which endure" (Rose 1992: 205). Synchrony is a most useful concept, in that the Dreamings all exists all the time, existing as Stanner's "everywhen." It is the land that links the two kinds of time, of the Dreaming and of the ordinary, visible world. As Rose puts it, "Both Dreaming and ordinary exist in real, named, localized space. Both are grounded in the earth, and both ultimately derive their life from the earth" (1992: 205). Again, and in yet another way, we find in Aboriginal culture an assimilation of time to space, to the land, and to the earth.

The experience of long duration is linked to the Aboriginal notions of the soul. In Aboriginal belief systems, souls can be made only out of souls. This means that those who are born must be forms of those who have been. Belief in the immortality of the soul provides an explanation for the perpetuity of the life of the clan and society. As Durkheim ([1912] 1965) wrote, "Individuals die, but the clan survives. So the forces that give it life must have the same perpetuity. Now these forces are the souls which animate human bodies; for it is in them and through them that the group is realized" (304). "For this reason," Durkheim concluded, "It is ... necessary that in enduring, they remain always the same; for as the clan always keeps its characteristic appearance, the spiritual substance out of which it is made must be thought of as qualitatively invariable" (*ibid.*). While the group, the clan, is not immortal in the absolute sense, it does endure far longer than does the individual. Even death, for the Aborigine, is an affirmation of what Swain calls "the enduringness of places" (1993: 45). While death to Westerners is an end to, or a miraculous saving of, personality, for Aboriginal people it is instead a statement of the spiritual continuity of the Ancestral places.

Myers (1986: 69), in his analysis of Pintupi Aborigines' concept of the Dreaming, sees this worldview as transcending the immediate and the present. The concept dichotomizes the world into that which is *yuti* ("visible") and that which is *tjukurrpa*, where the latter lies outside of human affairs and constitutes an *enduring*, primary reality. This construction occurs in space, on the landscape, where it creates places with enduring identity and relationships to other sites. Thus while the physical evidence of gigantic creator beings has been permanently embedded in the landscape, as artifacts of the inner world of *tjukurrpa*, the inner world itself is invisible. From a temporal standpoint, the spirit-world is potentially immanent in even the most inconspicuous things in the world, such as winds, plants, and animals. Thus even a humble snake could be possessed by a spirit being seeking the womb of the hunter's wife. The Aborigines have, as Stanner (1979) suggests, "a metaphysical emphasis on *abidingness*. They place a very special value on things remaining unchangingly themselves, on keeping life to a routine which is known and trusted" (38). Stanner explains that for the Aborigines, absence of change and certainty of expectation are so highly valued that they

> ... are not simply a people "without a history": they are a people who have been able, in some sense, to "defeat" history, to become ahistorical in mood, outlook, and life. This is why, among them, the philosophy of assent, the glove, fits the hand of actual custom almost to perfection, and the forms of social life, the art, the ritual, and much else take on a wonderful symmetry. If we put ... four facts about the aboriginal together, (1) an immensely long span of time, (2) spent more or less in complete isolation, (3) in a fairly constant environment, (4) with an unprogressive material culture, we may perhaps see why sameness, absence of change, fixed routine, regularity, call it what you will, is a main dimension of their thought. (1979: 38)

Durkheim, in his *Elementary Forms*, discussed the sense of duration—and of assent and acceptance—as an aspect of Aboriginal temporality. He saw Aboriginal artists as uncreative because of their static, traditional designs and because their paintings are limited to just four colors—white, black, red, and yellow. He speculated ([1912] 1965: 246) that "The dispersed condition in which the society finds itself results in making its life uniform, languishing, and dull." While he was keenly aware of the emotional energy generated and acted out in formal rituals, a collective effervescence, he saw everyday Aboriginal life as monotony with most actions utterly predictable (see Liberman 1985: 24). Durkheim's view is devoid of any basis in reality. Relying as he did on the secondary analysis of ethnographies, he had little to rely on beyond his own assumption that the Aboriginal mind possess but a "rudimentary ... intelligence," providing "no means for supporting any luxury in the way of speculation" ([1912] 1965: 75). It is ironic that the conversational style of Aborigines, in which experience is objectified, would appear a key mechanism of his "mechanical solidarity" that he was able to describe only with fanciful gestures of Aboriginal thought and social action. Again, to cite Liberman: "The ease with which Aboriginal gatherings objectify their experience is an identifying feature of Aboriginal social life and is related to the production of group solidarity" (1985: 44).

Perhaps the most obvious way to access the meaning of long duration and endurance in the everyday world of the Aborigine is through an analysis of the Aborigines' use of language in the creation of shared meanings. Liberman (1985) has made a superlative contribution with his analysis of tape-recorded and transcribed conversations of tribal-living, central-desert Aborigines. He observes that priority is given to a group structure rather than to individual initiatives and displays of personal agency. But within this group structure there is real intellect and real creativity. Liberman writes: "Within that group structure, Aboriginal people are highly creative in maintaining social harmony and the consensus which accompanies the congeniality they value so highly. Aborigines are incredibly artful in the skills of ordinary interaction—they rarely miss so much as a wink" (1985: 24). Their interactions are almost never characterized by competing proposals or hypotheses; they are, rather, based on a subtle sense of emotion. As Liberman notes, "Free of abundant plans and preconceived notions about what the results of a conference should be, they are open for the slightest direction of movement in the sentiments of an ensemble. What is more, they take great pleasure in engineering convivial social interactions" (1985: 25).

More to the point, Liberman directly addresses the Aborigines' experience of the long duration of their meetings and the enervation they experience in reaching consensus. Eager to elaborate on his sociology of the emotions, Durkheim ([1912] 1965: 405, 428) repeatedly spoke of the collective and joyful effervescence that results from participation in formal rituals. If, however, we broaden the concept of rituals to include everyday social activities, then Durkheim's generalization finds its limit because, as Liberman (1985) observes, in everyday interaction rituals, "the production of congenial fellowship is consummated not in collective effervescence but rather in a collective silence which makes further discussion excessive," in which the silence "is experienced as a sort of quiet bath in the fellow-feeling of the moment" (109). To this, it is helpful to refer to Maddock's (1972: 185) evaluation of the Aborigines' civilization as providing both freedom from oppression and freedom for positive and human social relationships, insofar as Aboriginal society includes mutual respect between societal members, a humanitarian anarchy, and egalitarian political relationships.

Duration for the Aborigines also has a feature of fixity and closure. The essential mechanism is consensus. Aborigines are reluctant to take action until all of, or nearly all of, the community or group members are openly showing their agreement. Ideas in Aboriginal conversations are developed cautiously, bit by bit, with no single person playing a strong leadership role. Durations are fixed not quantitatively, by the use of clock time, but rather qualitatively, with the moment of closure recognized by the attainment of a consensus. Liberman (1985: 64) provides an example. The following conversation takes place as a group of Aborigines is gaining consensus in closing a meeting. Aboriginal interlocutors utilize a serial order of formulating ideas in which a single possibility is cautiously and slowly developed as a consensus.

> APa Yes, that's all. Finished.
> APb Fine, we'll get another one.
> APc Yes.
> APd Good.
> APa That's the meeting.
> APe [whispered] I agree with what he's been saying, this old fellow.
> APf Yes, so it will be like that.
> APg Yes, that's all. Finished.
>
> APh Finished.
> APb The meeting is over.
> APg Finished, once more.
> APb So yes, that's [the] lot.
> APa That's all, that's [the] lot.
> APc Yes, finished.
> APd All done.
> KL Finished?
> APi Yes.

As a final note, the experience of long duration is reflected in veneration of the elders, in individuals who have personally endured. The highest status in traditional Aboriginal society is reserved for elders who have gradually gained wisdom and a deep knowledge of their own Law, culture, and customs. Gray hair is a status symbol, and to call another an "old fellow" is an expression of respect. For the Aborigines, status and prestige come through the duration of their experience and the knowledge that they have developed and reproduced. An Aboriginal man with a section of his beard turned gray will typically let that part grow longer than the rest. Status and prestige in Western society, in contrast, tends to be bestowed on the young, and there has emerged an anti-aging industry aimed at preserving the appearance of youthfulness through hair dyeing, Botox™ applications, face-lifts, and more, so that old age has come to be seen, by some, not as a natural process accepted by an authentic temporality but rather as a pathology.

Chapter 5

Ordinary-Linear Time Consciousness

In the modern world, just as in pretechnological societies, there exist temporal patterns and regularities. Time is created in the interactions of people and things; it is the crucial dimension of all human social interaction (Gosden 1994: 122). As Durkheim recognized, both natural and biological rhythms are used on the cultural level. We have seen that an important kind of time consciousness, characteristic of the Australian Aborigines, is based on an overall pattern or gestalt. In this chapter we shall see that *ordinary-linear* time consciousness can be adequately described by seven features that are structurally the opposite of the seven aspects of the patterned-cyclical kind of time consciousness. By imagining ordinary-linear time to be a line, we find that its seven features are descriptive of a line consisting of an infinite collection of ordered points, each point corresponding to a real number: it is a single dimension (L1), partitionable into the past, a point representing the present and the future: whereas cyclical time fuses the past and the present, in linear time, it is primarily the future that is fused with the present (L2); continuous (L3), measured by clocks and calendars (L4); used to mark events, which can be ranked on the dimension in terms of priority, simultaneity, and posteriority (L5); numerical/quantitative, with an interpretable zero point (L6); and interpretable as an infinite sequence of moments (L7). The two models, of patterned-cyclical and ordinary-linear time, are summarized in table 5.1. The mode of thinking associated with ordinary, linear time, as we shall see in the next chapter, is propositional, logical-analytic, and linear, and it has as its infrastructure the left hemisphere of the brain.

TABLE 5.1 Conceptual models of ordinary-linear and patterned-cyclical forms of time consciousness

Ordinary-Linear Time	*Patterned-Cyclical Time*
L1. Linear, time as a single dimension	P1. Dualistic, split into two levels of reality
L2. Separation of past, present, and future	P2. Fusion of past, present, and future
L3. Regularity, continuity, and homogeneity	P3. Irregularity, discontinuity, and heterogeneity
L4. Clock orientation and calendar orientation	P4. Event orientation and nature-based orientation
L5. Diachronic ordering of events: priority, simultaneity, posteriority	P5. Synchronic ordering of events: cyclical, patterned, oscillatory
L6. Quantitative: numerical measurement; an invariant anchor or a zero point	P6. Qualitative: nonnumerical measurement; now the anchor point
L7. The experience of time as a series of fleeting moments	P7. The experience of long duration

Linear Time As a One-Dimensional Vector

Heidegger ([1927] 1962) is best remembered for his notion of a primordial temporality primarily determined out of the future, but his description of "ordinary," linear clock time is of great value in its own right. His "ordinary" time first of all possesses the significations "past," "present," and "future." In our everyday concerns with time, we say that something is to happen "then," that something else must be considered "beforehand," that something we failed to do "on a former occasion" must be made up for "now." Every "then" is, as such, a "then, when ...," every "on a former occasion" is an "on a former occasion, when," and every "now" is a "now, that." This understanding of time as a continuously enduring sequence of pure "nows" is immediately intelligible and recognizable. Any such sequence can be modeled as a straight line, the variable t, and as a single dimension defined as the array of all real numbers that goes together with the x, y and z of Cartesian spatial coordinates.

Past, Present, and Future

In chapter 3 we saw a fusion of the past and present in the Aborigines' religious practices; it was argued that this "fusion" or "interpenetration" of the present and the past is a defining aspect of patterned-cyclical time consciousness. By viewing the future as part of the present, inner reality, there exists an ontological principle of order in which humankind, nature, and society, along with the past, present, and future, are seen as an enduring totality or gestalt. What is grasped as a whole is not any set of discrete time units but rather time experienced as a flux (Kern 1983: 24). What Bergson ([1910] 1960, [1911] 1975) called "living in the *durée*" corresponds to one important part of the definition of patterned-cyclical time consciousness, namely, a fusion of the three tenses.

But ordinary-linear time consciousness is in a sense the opposite of Bergsonian fluidity insofar as it articulates classificatory distinctions between the tenses of time, past, present, and future, which refer to the transitional nature of time. The past, present, and future can be thought of as a "frame" through which mental life flows. Wyndham Lewis (1927) notes that Bergson saw the three tenses as "penetrating" and "merging" but suggests that one also can "enjoy the opposite picture of them standing apart—the wind blowing between them and the air circulating freely in and out of them: much as he enjoys the 'indistinct', we value the distinct, the geometric, the universal" (120, 428). Thus the opposite of fusing the past, present, and future is to keep them separate so the linear dimension is partitioned into the past segment, the present point, and the future segment.

This linear conceptualization of time is socially institutionalized on a global scale. It is deeply taken for granted and widely considered self-evident. The past and the future revolve around the present but are not actual and therefore are not seen as real. As Heine (1985) writes, "The past comes before the present and has been fully exhausted when the present takes place; and the future which has not yet arrived is equally empty of actuality" (2). We refer to the "before and after," the "beginning and end," only relative to the presumably stable present. Such terms reveal an attachment to a "now" that partitions time into three isolated and *independent* tenses. Scholars of time have, in general, not gone so far as to argue that linear time is "false" or lacking in utility. The problem arises when this fabricated or derivative kind of time is taken to be absolutely real, corresponding to the psychological experience of time. Under this condition, it becomes possible to feel victimized and controlled by time, seeing it as necessary in order to do things, and seeing it as easily slipping away, as something that is somehow in short supply. However, any effort to fight against, to somehow overcome, this movement of time, at least according to Heidegger, is apt to take us even farther from an *authentic* experience of temporality. Thus the problem with linear, derivative time is that it tends to conceal more primordial forms of temporal experience.

Homogeneity, Regularity, and Continuity

With the development of urbanization and mercantile and industrial capitalism, and with an intensified circulation of money, the emphasis of time consciousness shifted from the cycles and rhythms of nature and related patterns of culture to the idea of forward progress and the regular accumulation of nonrepeatable moments (Whitrow 1972). The tempo of modern life has accelerated an artificial time environment punctuated by mechanical contrivances and electronic impulses. Time came to exist as a flowing, thoroughly uniform, homogeneous, and successive being.

In modern society there is great regularity in the biotemporal and sociotemporal levels of social organization. This regularity can be seen in an examination of the consequences of temporal *ir*regularity. Zerubavel (1981: 22) takes the methodological approach of looking at a pathology in order to understand the nonpathological. The pathology in question involves a cognitive incongruence between the social figure and its temporal ground. The primary emotional reaction to such a situation of spatiality is one of surprise.[1] Manifestation of surprise demonstrates the existence of prior expectations in turn indicating an expected regularity. Zerubavel argues that temporally regular patterns—largely taken for granted and thereby ignored—regulate much of our social life and make it "understandable that we would have certain expectations regarding the duration, sequential ordering, temporal location, and rate of recurrence of many events in our environment" (1981: 23). The figures we perceive have meaning only if they occur in their ordinary and usual background expectancies. The temporal regularities of the everyday world are important among the background expectancies that are the basis of the "normalcy" of our social environment. The regularity and predictability of our temporal structures of social life are responsible for the establishment of a solid temporal ground against which the occurrence and the presencing of certain people and objects pass as "normal." When an unexpected event occurs, or an unexpected person appears outside of her or his expected temporal niche, there is a disruption of the implicit and taken-for-granted figure-ground configuration of activity presupposed in our normal social environment. The result, Zerubavel suggests, is "a chaotic incongruity between figure and ground, which entails the loss of a meaningful way of anchoring our cognitive experiences" (1981: 21). The act of leaving a party after twenty minutes or of calling a casual acquaintance several times every day leads to the discomfort and surprise of the host and acquaintance, violating implicit norms of sociotemporal order. Thus the norm of regularity is brought into relief by studying the pathologies of disturbances. The high level of intolerance toward temporal anomalies shows the rigidity of the temporal structures of social life and also serves to sustain this rigidity.

Clock, Schedule, Timetable, and Calendar Orientation

Many civilizations of ancient people developed a capability to reckon time spans for the time of day, month, and year. Water clocks existed in ancient Egypt, perhaps as early as the fifteenth century B.C. (Adam 1995: 27). Archaeological evidence of lunar time reckoning possibly goes back some 30,000 years (Marshack 1972). The beginnings of the concept of time derive from the human need to fix the location and duration of happenings in the succession of events. Events can be partitioned into two kinds, those that repeat and those that do not repeat. Time measurement came about as an effort to quantify a specific type of regularly occurring natural processes as a standard sequence. Clocks, together with calendars, provide the means of measuring and thereby regulating public time. Clocks make possible the comparison of the duration and speed of perceptible processes separated in time.

The reified, abstract, machine time that we associate with the Industrial Revolution and the modern world historically depended upon the development and use of accurate clocks, mechanical devices designed using the laws of classical mechanics that produce sequences of regularly occurring, precise, and observable physical events having identical durations. These mechanically produced events are characterized by idealized invariance, artificial precision, and context independence (Adam 1990: 50–55, 1995: 24). By clocks and calendars, the duration of events and processes can be provided with numerical comparisons according to a hierarchy of decontextualized and arbitrary units—picoseconds, nanoseconds, seconds, minutes, hours, days, weeks, years, decades, centuries, and millennia. The repetition of timing events is cyclical only in the narrow sense that all cycles have identical temporal durations. Clocks thereby provide for the measurement of time, for the "how long" and the "when," the "from when till when." Adam observes that "[t]he abstract, quantitative, spatialized time of clocks and calendars forms only one aspect of the complex of meanings associated with the time 'when', the time that forms a parameter of our existence and locates us *in* time" (1995: 21). Clocks show us the "now" but not the past or the future. If a clock is used to determine the point in time when a future event will take place, it is not the future that is measured, it is the "how long" one has to wait until the now intended occurs.

With the fourteenth-century invention of the mechanical clock, the time of the day came to be divided into sequences, intervals, and durations, all imagined as a straight line along which events can be arranged. In the modern world, punctuality became a social necessity as clocks and watches began to proliferate, becoming commonplace in Britain in the 1790s. Well into the modern era, different cities within the same country had differing local times, a problem for public transportation. In New Zealand, for example, it was not until 1868 that the North and South Islands announced that the time was to be standardized corresponding to longitude 172° 30' East of Greenwich. The synchronization of local times was accomplished in the 1840s in England, necessitated by the introduction of railway transportation, and in the United States thirty years later. The protracted process of accurately measuring public time, which has existed in simpler form in the Greek city-states, was permanently resolved with the worldwide standardization of time that came out of the Paris International Conference on Time of 1912. This problem of coordinating travel within any geographic space was solved by the invention of the telegraph, which left virtually no interval of time between widely separated places, as information could now be transmitted at the speed of light. The advent of globally standardized time made possible a linkage of time and space in the calibration of the movements of people, goods, and information in public life. Railroad companies first instituted standardized time to overcome the confusion resulting from multiple local times. For example, around 1870, a traveler from Washington, D.C., to San Francisco, California, would, in order to maintain correct time, have to set his or her watch to over 200 local times. The railroads imposed a uniform time-zone system in the United States in 1883, which increased their efficiency and profitability (Kern 1983: 12). The introduction of world time was a prerequisite for a global economy in which information on prices, supplies, demands, commodities, and capital can be rapidly transported or transferred. Standardized time also facilitated global warfare, as in 1914, when the world went to war according to mobilization timetables. The Great War imposed homogeneous time. In the prewar years, watches had been thought of as feminine and unmanly, but during the war they became standard military equipment, thereby losing this connotation.

The modern calendrical system was in its origin a technical innovation, the result of numerous small changes and adjustments built into a pattern that gained nearly global acceptance. The calendrical system that had not changed for decades or centuries gradually came to be accepted tacitly and was taken for granted. Then, periods such as the second, minute, and hour came to be represented in social consciousness, incorrectly, as "natural" divisions of time. Hall ([1959] 1973) gives as an example the change in calendars in 1752 in England to the Gregorian version (the correction setting the time ahead fourteen days), which led to riots in the street, where outraged citizens shouted, "Give us back our fourteen days" (9).

Clocks, together with calendars and schedules, thus provide the means of measuring and thereby regulating and timing public life, including communications, industry, war, and everyday events. Clocks have come to represent time per se as the abstract measure of motion, a measurement methodology that has become incredibly sophisticated and accurate. Clocks are a means of cognitive orientation within the succession of physical, biological, and social processes.

Clock time is culturally transmitted by schools, "where bells and buzzers, clocks and calendars reign supreme" (Adam 1995: 59), and where all daily activities and long-term objectives are timed in a normative way. Students, teachers, and administrators are bound together by "a common schedule within which their respective activities are structured, paced, timed, sequenced, and prioritized" (Adam 1995: 61). The overall timetable of the school provides all role incumbents with regular, dependable routines within which there is a careful scheduling of various activities—attending to selected academic topics, examining, assessing, managing, engaging in sport, eating, reading, and so forth (Delamont and Galton 1986). Educational outcomes—the accumulation of information and knowledge, are brought about by educational processes that presuppose a time consciousness in which the occurrence of events is timed according to clocks, schedules, and timetables, as they interface with, and are coordinated with, personal commitments of time and energy. Schedules and calendars are thus "constituted on the basis of individual and collective histories and futures which, in turn, have a central bearing on any one moment of time generated by the group" (Adam 1995: 67).

Time in the school is hierarchically organized, with activities such as breaks, arrivals, registrations, and participatory events meticulously planned in the name of efficiency and effectiveness in a way that ignores individual and local differences and preferences in the amount of clock time allocated for the completion of tasks. Control of timetables and schedules demonstrates power and influence (Schwartz 1979). Clock time and calendar oriented time are fully quantitative, making possible their segmentation and compartmentalization and their use in complex organizations as finite, measured resources to be allocated, appropriated, partitioned, budgeted, prioritized, controlled, and managed (Adam 1995: 62–63).

Individual members of Western (and more generally modern) societies manifest a sensitive consciousness of clock time. Emphasis is placed on promptness and the scheduling of planned future events. Persons who are systematically late for scheduled events come to be viewed as irresponsible. Promises to meet deadlines and appointments are taken seriously. There are substantial penalties for not keeping commitments, contracts, and agreements in agreed-upon time frames. Once set, a schedule becomes marginally sacred insofar as it and its boundaries, while of course subject to negotiation, must be treated with respect.

The historical change in time consciousness from event based to clock based corresponds not only to industrialization and the regulation of everyday life but also with the emergence of modern science, especially classical physics. The invention of the clock played an important role in the development of a mechanistic conceptualization of nature. For classical mechanics, and more generally for the Newtonian worldview, the symbol of nature was the clock (Adam 1990: 50–68; Prigogine and Stengers 1984: 111). Central to Newton's ([1729] 1934: 6) theory was the concept, defined in 1687, of Absolute Time, which of itself, and from its own nature, flows continuously[2] without regard to anything external. Absolute Time was viewed as a linear medium that exists independently of human consciousness. Consistent with his calculus, Newton saw time as an integral of infinitesimally small, discreet units. Clocks produced an audible manifestation of the atomistic nature of time. With the aid of his concept of time and space, motion was conceived of as the movement of objects in an abstract space of three dimensions. Thus time came to be objectified through being linked to scientific knowledge. Once so objectified, time "lends itself to rational calculation and precise counting" (Ferrarotti 1990: 87). In the modern world of science, as Dorothy Lee (1949) puts it, "… chronological sequence is of vital importance, largely because we are interested not so much in the event itself, but rather in its place within a *related* series of events; we look for its antecedents and its consequences. We are concerned with the causal or telic relationships between events" (406, emphasis in original).

This method of timing events provided for the regulation of informal and especially formal social activities. All social activities draw on a fund of knowledge that serves each generation as the human means of orientation to the world (Elias 1993). The invention of clocks enhanced the abilities of people to coordinate their social activities and situate related events and activities. Elias writes that even before Galileo, "Time was above all a means of orientation in the social world, of regulating the communal life of human beings" (1993: 3). Agreed-upon clock time exists as a sophisticated social construction for the coordination of body, self, society, and nature. The impact of clock time is not restricted to the scientific and economic spheres of social life, for all of modern social organization is temporally regulated by the clock, the calendar, and the schedule.

The elaborated sociotemporal ordering of modern social life developed out of the necessity first of urbanization and second of the Industrial Revolution. Rotenberg (1992), in his study of metropolitan Vienna from the mid-nineteenth century to the present, shows that public schedules existed before industrialization but gained importance as heightened economic activity led to population increases and as workplaces became more distant from workers' dwellings. The larger the city and its population, Rotenberg found, the greater the mandate for schedules and timetables to coordinate with the circulation of residents in their various domestic, leisure, work, and school activities. Schedules, as Zerubavel (1991) notes, are "[p]robably most responsible for the establishment and maintenance of temporal regularity in our daily lives" (xiii). Calendrical time facilitates the regulation of important aspects of social existence—the setting of holidays and other significant dates, the temporal determination of contracts, the scheduling of work and leisure activities, and so forth. The social structure of modern society imposes on its members an inescapable network of temporal definitions and expectations, contributing to a personality structure that has an acute and a disciplined sensitivity to time (Elias 1993: 7).

Diachronic Ordering of Events: The Before and the After

Once time is viewed as a linear dimension, events can be ordered as points on a continuum. For any two events, A and B, occurring as specific points in time, there are three possibilities: first, A and B can occur simultaneously; second, A can occur prior to the occurrence of B; and third, A can occur after the occurrence of B. Since Aristotle, the events that occur in the world are prior or posterior, with time an attribute of the changes and interactions of physical entities.[3] That the meaning of the concept "simultaneity" is not culturally universal is demonstrated by Hall's account of the inhabitants of the atoll of Truk in the Southwest Pacific. The time consciousness of the Truks would appear to deny the simultaneity of events physically separated. Hall ([1959] 1973) writes, "Though the Truk islanders carry the accumulated burden of time past on their shoulders, they show an almost total inability to grasp the notion that two events can take place at the same time when they are any distance apart" (16).

Certain kinds of events can be seen as occurring at single points of time, but other events unfold over a period of time. Thus we can speak of the duration, period, or temporal thickness of an event and of the duration or period between events. This is not a qualitative experience of long duration but rather a quantitative measure of an exact or estimated duration. The stages of time stand in a relation of earlier and later to one another. The kind of time that we encounter in everydayness, that has been elaborated in physical science, provides for the measurement of space-time relations. Just as *there is no such thing as time*, there is also *no such thing as space*. Aristotle long ago correctly argued that there is no absolute space but rather a relative space, which exists merely by way of the energy and the bodies that that space contains. But as Spengler (1926) argued, Space is fundamentally a *concept*, and "*Time is a counter-conception (Gegenbegriff) to Space*, arising out of Space" (126, emphasis in original).

Quantitative Chronology

The habit of linking "time" to "space" made the concept of time a variable to be estimated and measured. The result is that space and time are essentially related to calculation. Linear time is abstract and mechanical, divisible into small, interchangeable units that can be fixed, arranged, scheduled, and parceled. "Linear" time is quantitative time. Time becomes a "chronology," calibrated into a standard gauge against which we associate events. With the historical development and widespread use of clocks, the making public of ordinary, everyday time was greatly enhanced. It is no longer necessary to glance at the sun and ascertain its position in order to approximate the time of the day, for a glance at one's watch makes it possible to tell the time directly and exactly. This "reading off" of the time signifies, one might say, "It is now such and such an hour and so many minutes; now is the time for …," or, "There is still time enough now until …" In the doing of ordinary time, we look at clocks and thereby regulate ourselves "according to the time." The "now" in all of these examples "… can be understood and interpreted in its full structural content of dateability, spannedness, publicness, and worldhood" (Heidegger [1927] 1962: 469). "The measurement of time," Heidegger wrote, "gives it a marked public character, so that only in this way does what we generally call "the time" become well known" ([1927] 1962: 471).

Through rational use of time in economic activity, time becomes a quantity to be used, allocated, and exchanged (Adam 1990: 113). The quantitative time of Western society is in one way similar to currency, the value of which fluctuates with supply and demand. Capitalist work discipline abstracts both work and time from contexts and meanings, imposing an independent, metric dynamic. Nowotny (1975) sees this abstraction from meaningful context as having been achieved only through the quantification of time, through the expression of time in numerical values. Thompson (1967) saw that lived time, the substance of being and social life, has become extended by a second notion of time, time as an abstract quantity of pure duration, this quantitative time being freely exchangeable with all other times and serving as an abstract medium of exchange. Adam (1995) suggests that time, despite its commodification, can be conceptualized as "a spatial quantity by which time is measured as distance" and exists as "resource, as money, and as clock-based rhythm" (75). Linear time functions "as a habituating environment, as preparation for an industrial way of life" (Adam 1995: 76). According to Luhmann (1982), this abstract, quantitative measure of time, in combination with objectified time frames or chronologies, is a recent historical event that marks a change from a fusion of events and time—time *in* events that, by objectification and abstraction, leads to a linear conceptualization of world time. The quantified timing of work has led to the timing of other spheres of everyday life. Most adult persons now carry watches, some day and night, so that numerous forms of behavior can become as regular as clockwork (Woodcock 1944).

Knowing one's chronological age, by which the duration of one's life becomes a variable with which age can be precisely measured in units of years, was not possible until the advent of the calendar. Knowing one's age, as an exact sum, was not common in Europe until the eighteenth century. The calendar provides a frame of reference consisting of a sequence of recurring divisions socially standardized as days, months, and years. This makes possible the fixing of one's birth to a certain day in a certain year. It also provides for the comparison of the numbers of calendar years any two people have lived. In modern societies it is taken for granted that every normal adult person knows his or her age.

In the modern world, people count the years of their age, celebrate the day of their birth each year, and are apt to have the dates of their birth and death written on their graves. Important milestones of life are marked by age, and determined by age as well: a child starts school at age five, is a "teenager" from ages thirteen to nineteen, and gains the privileges of adulthood on birthdays that range from ages sixteen to twenty-one, such as voting, being licensed to drive an automobile, and being allowed to consume alcohol. The calendar makes it possible to compare the ages of societal members known personally or indirectly through media exposure. In order to compare the ages of any two persons, it is necessary to have as a frame of reference another sequence with recurrent division, the length of these divisions being socially standardized. It is, by definition, the calendar that made possible this frame of reference (Elias 1993: 6). The calendar sets forth an unrepeatable succession of numbered days and years symbolically representing the unrepeatable succession of social and natural events. It is the articulation of events with times that enables societal members to orient to the social and natural worlds.

The Experience of Time As Fleeting and Flying By

While in patterned-cyclical time there is an experience of long duration, of accumulation of experience, the opposite of this experience is time as transient, as fleeting, as flying by. What is characteristic of ordinary, linear, "time," of clock time, is the fact that it is a pure sequence of "nows," an infinite sequence without beginning or end. Heidegger noted that this ordinary understanding of time "shows itself as a sequence of 'nows' which are constantly 'present at hand', simultaneously passing away and coming along" ([1927] 1962: 474). Time is thus understood as a succession, as a "flowing stream" of "nows," as the "course of time."

When we say that "time flies" or that "time is passing," we are apt to inadvertently articulate a sense of subjugation of societal members *by* time, that the passing of time is an inevitable, a ceaseless, a relentless, and an unstoppable movement. The sequence of nows cannot be prevented, delayed, or reversed, so there is never enough time. This conventional, ordinary view of time was first systematized by Aristotle, with his definition of time as the substantive duration of the now in relation to the receding no longer now and the approaching not yet now. Aristotle's time was defined by its countability, as he said: "Time, then, is the number of movement in respect of before and after" (1936: 387). The present instants are seen as a series of comings and goings that marches forward in a stable way that is uncontrollable and unconnected to the past and the future.

Chapter 6

Patterned-Cyclical and Ordinary-Linear Time and the Two Sides of the Brain

We have already established seven-part definitions of both patterned-cyclical and ordinary-linear time consciousness. It is a fundamental postulate of the present theory that these two broad cognitive structures are expressions of still more general modes of information processing, the gestalt-synthetic processing capability of the right side of the brain and the logical-analytic processing of the left side of the brain. This claim will require a short review of the neuroscientific literature on dual-brain theory, the theory of cerebral lateralization of the higher mental functions. Prior to this discussion, we will deal briefly with two topics pertaining to linear time, the postmodernist critique of linearity and the psychophysiology of duration estimation.

THE POSTMODERNIST CRITIQUE OF LINEARITY

In sociology and related social-scientific disciplines, and in social philosophy, postmodernist scholars have developed a conceptualization of contemporary society as decreasingly linear and less based on rational choice and rational behavior, arguing instead for a worldview that sees social order as increasingly chaotic, fragmented, and discontinuous. Given these currents of thought, one might pause to wonder whether the venerable distinction between ordinary-linear and nonlinear, cyclical time represents only an otiose dualism appropriate to classical eighteenth- and nineteenth-century thought. Theorists of postmodernism are indeed apt to see linear time as obsolescent, replaced with the reality of a fragmented, deconstructed time.

Process, development, and history, in the postmodernist vision, are seen not as continuous and linear but rather as characterized by abrupt changes, catastrophes, breaks, and erasures (Masuzawa 1993: 161).

In spite of these arguments, time perception is, and will remain, linear. The exact measurement of physical time will always be important to engineering, technology, physical science, biology, social science, and other fields of human inquiry. Moreover, clocks and calendars are so deeply and permanently embedded in the structuration of everyday social life that their removal would result in significant social disorder. Even where the focus of investigation is on living systems, nonlinearity can only be inferred on the basis of comparison to linearity. In mathematics, the concept of a line as a single dimension will always be important.

The Psychophysiology of Duration Estimation

The perception of duration is linear on a psychophysiological level that will hardly be rendered obsolete by scientific, cultural, and societal developments. The reason for this is that an accurate estimation of duration has survival value for human beings. While our responses to pressure and smell vary nonlinearly with the intensities of the stimuli, our estimation of duration is, and will remain, nearly linear. Stevens (1975) found that psychological estimations (e.g., of the magnitude of an electric shock) increase as a power function in accordance with Plateau's (1872) Law. That is, perceived magnitude (Ψ) is proportional to physical magnitude (M) raised to some power (m), or $\Psi = kM^m$, where k is a curve-fitting constant determined by choice of units. Stevens (1960) and others have tested many kinds of stimuli under this model and found that the exponents range from 0.33 for brightness of white light to 3.5 for electric shock. Exponents less than unity yield negative exponential functions; exponents greater than unity, exponential functions; and exponents close to unity, linear functions.

For sensory channels that are highly buffered, such as audition and vision, exponents of the power function are low. For sensory channels for which the effective range is smaller, such as temperature and pressure, the exponents are higher, meaning the effects of physical changes are magnified. Small exponents mean that the transduction properties of the sensory receptors reflect "compression," which helps the nervous system resist being overwhelmed by sudden-stimulus energy changes (Dember and Warm ([1960] 1979: 93–94). The different operating characteristics for different kinds of stimuli are shown in figure 6.1.

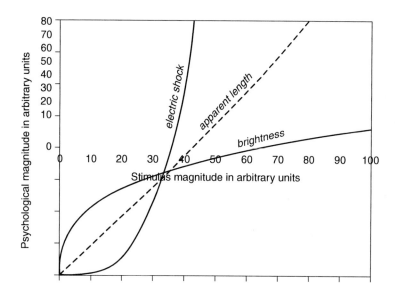

FIGURE 6.1   The psychophysiological power law: equal stimulus rations produce equal subjective ratios. The apparent magnitudes of electric shock, length, and brightness follow different curves of growth, because their power law exponents are 3.5, 1.1, and 0.33, respectively. The curve is concave upward if the exponent is greater than 1, linear if the exponent is about 1, and concave downward if the exponent is less than 1. Estimates of apparent duration, such as apparent length, result in nearly linear functions.
Source: Stevens 1961, figure 4, p. 11.

The kinds of stimuli that come closest to unity are distance and duration. For the duration of white noise, stimulus $m = 1.1$ (cold temperature on arm also has $m = 1.0$ but for warm on arm, $m = 1.6$) and repetition rate $m = 1.0$, for light, sound, touch, and shock (see Dember and Warm [1960] 1979: 91–93). When the luminance of a spot of light in a dark field is doubled, the estimate of apparent brightness increases by approximately 25 percent. But if a current of electricity through the fingers is doubled, then the sensation of shock increases about tenfold. For most people, a length of 100 centimeters looks about twice as long as a length of 50 centimeters. For time-stimulus estimation, the estimated duration of time of exposure of the stimulus is nearly a linear function of actual stimulus duration (Stevens 1961: 10–12): Simply put, this means that our estimate of the amount of time that has gone by is directly proportional to the amount of time that has gone by. This generalization holds true in controlled laboratory conditions, but it is well known that time is experienced as slowing down under two conditions: when there is nothing of interest going on and one is bored (e.g., while waiting for a pot of water to boil), or when one is engaged in a process that requires one's full attention (Flaherty 1999) (e.g., as one is falling off of a ladder or attempting to catch a football under duress).

Temporal Experience and Cerebral Lateralization

Before turning to the topic of cerebral lateralization, per se, it is helpful to place this topic in a broader context of brain organization. Three principal functional units of the brain are involved in all kinds of mental activity. As explained by Luria (1973: 43–101), they are the following:

(1) A unit for regulating tone and wakefulness, the reticular activating formation (Magoun 1963), which is obviously a prerequisite not only for all consciousness of time but also for all other higher mental functions;

(2) A unit for obtaining, processing, and storing information from the outside world, which is located in the posterior regions of the neocortex, on the convex surfaces of the three posterior hemispheres, including the visual (occipital), auditory (temporal), and general sensory (parietal) lobes (see figure 6.2). The primary cortical areas, responsible for somatotopical projection, are not lateralized, but lateralization of function is

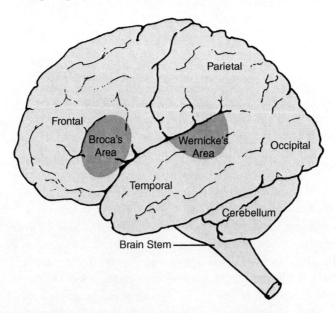

FIGURE 6.2 Side view of the left hemisphere of the human brain, showing approximate locations of the four lobes, with the two language areas shaded.
Source: Hellige 1993, figure 1.1, p. 7. Reprinted with permission of the publisher.

found in the secondary projection-association areas, responsible for coding primary projections into functional organization, and especially in the tertiary zones, responsible for the production of supramodal, symbolic schemes, the basis for complex forms of gnostic activity, culminating in two complementary modes of information processing, the gestalt-synthetic and the logical-analytic. Thus our two modes of time consciousness, the patterned-cyclical and the ordinary-linear, depend on the work of the second functional unit of the brain, the specializations of the posterior right and left cerebral hemispheres;

(3) A unit for programming, regulating, and verifying mental activity. The human mind, to act effectively, not only needs to process information but also to develop intentions, to form plans and programs of action, to evaluate and modify performance directed to these plans, and to compare the actual effects of actions with their original intentions. The basis of this unit is the frontal lobes, the anterior portion of the brain in front of the precentral gyrus, which occupies about one-fourth to one-third of the total mass of the brain. The prefrontal regions of the brain are tertiary structures intimately connected to nearly every other zone of the cortex, and they constitute "a superstructure above all other parts of the cerebral cortex, so that they perform a far more universal function of general regulation of behavior than performed by the ... tertiary areas of the second functional unity" (Luria 1973: 89, emphasis deleted). It will be shown, in the next chapter, that two *other* kinds of time consciousness—the episodic-futural and the immediate-participatory—involve the frontal lobes and the *primary* association areas of the posterior cortex, which carry out very general modes of information processing that Pribram (1981) has called "episodic" and "participatory," respectively.

Time, being an abstraction, cannot itself be an object of perception. It is, rather, an attribute of our mental construction of events. On this basis, it is not only possible but also necessary to relate the experience of time to brainwork, beginning with Luria's second functional system, and to the two cerebral hemispheres. In the well-lateralized, right-handed adult, it is usually the case that the left and right hemispheres (LH, RH) are specialized for different modes of information processing. A great number of dichotomies have been advanced to characterize the cerebral lateralization of higher mental functions. Luria (1973) views the LH as the language hemisphere, arguing that verbal versus nonverbal is the primary determinant of lateral asymmetry. Corballis (1991) similarly claims that the LH's language mechanisms are the main aspect of lateral asymmetry. He contends that there has been a great leap forward in brain evolution, separating the human from other primates, this evolutionary discontinuity coming about with the advent of the "generative mind," whose functions are housed only in the LH. This putative "general assembling device" (GAD), he implies, might have resulted from some miraculous "nonmaterial intervention" (Corballis 1991: 16). Greenfield (1991; also see peer commentary and reply), however, presents an impressive argument, based on solid experimental evidence, that during the first two years of human life a common neural substrate (Broca's area) underlies the hierarchically organized, *sequentially ordered* information processing that is a prerequisite for the behaviors associated both with language and tool use. She hypothesizes that this biological foundation of language was present in primates before the hominids and the great apes diverged. Support for this view comes from the discovery of a Broca's area homologue and associated neural circuits in contemporary primates. Further, in chimpanzees, there is an identical constraint on hierarchical complexity in both tool use and symbol combinations, with performance matching that of the two-year-old human, who still lacks the connectivity of neural modules necessary for the attainment of these two cognitive abilities.

A vast body of experimental studies, using visual, auditory, and tactile stimuli, has conclusively demonstrated that it is not the stimulus material per se but rather the cognitive processing induced by the task at hand that determines lateral asymmetries in cognitive functioning.

The LH operates in a logical, an analytic, and a propositional mode, with an emphasis on successive, temporal, sequential, and linear processing, and is responsible for calculation, arithmetic operations, numerical cognition, and, in general, quantitative analysis (Geschwind and Galaburda 1987). The LH is specialized for tasks that require analysis of the internal structure of stimuli (Bever, Hurtig, and Handel 1976) and syntax in language use, both for relations between words (except those based on melody) and the meanings of words. The human use of complex tools, including language, requires thought that is sequential and time dependent and that uses syntactical mechanisms for generating new sequences according to grammatical rules. Thus language emerges as a means of communication and also serves as a cognitive and conceptual tool for categorizing the world (Bradshaw and Nettleton 1983).

Corballis (1991) sees the RH's greater involvement in the cognitive representation of the emotions and in a wide range of spatial, compositional, and visuoconstructive tasks as not distinctive of our species, suggesting that these relative differences are not the result of a RH mode of thought, nor as evidence that we can think effectively without language, but rather of LH *deficits* resulting from the crowding out of spatial reasoning in the LH by the GAD mechanism. This position is in opposition to Geschwind, who saw wide continuity across species in the left lateralization of communicative processes, but close to Nass and Gazzaniga (1987: 722), who liken the RH of the human to the mind of a chimpanzee or even a monkey. That language somehow is a gift from God might be a comforting thought to some, but there remains the possibility that the RHs of individual *Homo sapiens* such as Beethoven and Leonardo da Vinci, who had probable LH language difficulties (see Ehrenwald 1984), were more than monkeylike or chimplike in their cognitive functioning. Such instances of human genius based on thinking without language clearly refute the arguments of Corballis, Gazzinaga, and Nass. We hardly need to embrace a romantic view of the RH and its "creativity," or an inflationist view of human consciousness, to reach this conclusion.

The RH synthesizes information from the environment; it recognizes forms, things, and arrangements; it images, symbolizes, composes, and gestalts and it is involved in perceptual closure; and it looks inward to assess the state of the body in relationship to its activities, moods, thoughts, and situations and also looks outward to orient the body and mind in extended space, in the process bearing a primary responsibility for spatial reasoning, including the recognition of objects and things. Damage to the RH—which makes the workings of the mind more dependent on the preserved LH—typically leads to a preoccupation with details rather than the whole; a lack of overall organization; a variety of deficits in form recognition; problems in spatial relations and route finding; a reduced production of images, fantasies, and symbolizations; deficits of visual closure and other forms of gestalt completion; and deficits in the perception of relations between component parts and the whole in performing spatial transformations and manipulations of visual input. The RH is primarily responsible for forming spatial and cognitive models of our surroundings (Bogen 1969, 1977; Geschwind and Galaburda 1987).

The RH maintains *in the present* a spatial-cognitive model of our surroundings, a global awareness of the web of life, of the patterns, processes, and oscillations of nature and culture; it is a worldview, a dynamically interrelated basic cognitive structure that is collectively held and constantly updated. Given this evidence in support of the RH being capable of a gestalt-synthetic mode of information processing, it is reasonable to infer that the RH is much involved in the mental activities of cultural members who celebrate the sharing of paleosymbolic images expressing a shared worldview. There are two aspects to any worldview, and patterned-cyclical time consciousness is no exception. First, is its *content*, the description of which is the task of ethnography. The Aborigines were selected for the case study precisely because of the excellence of the ethnographic studies that have been carried out of their belief systems, cultural practices, mentality, and conceptualization of time and temporality. Second is its *structure*, the basic organization of thought shared with all human worldviews (see Kearney 1984: 3).

Levy-Agresti and Sperry (1968) introduced the terminology used in this book, seeing the RH as "gestalt-synthetic" and the LH as "logical-analytic" in their general modes of information processing. Bradshaw and Nettleton (1983) claim that their similar "holistic" versus "analytic" distinction considers the LH specialized in terms of its capacity for temporal order and sequencing. The preferential mediation of language by the LH, they argue, results not from a specialization for symbolic or phonological processes but rather from its specialization for "analytic" processing; the RH, in contrast, has a more immediate mode of apprehension, of the whole configuration, and of the patterns of interrelationship, while otherwise independent of the component features, which can be understood as a specialization for "synthetic" processing. Thus the LH would appear to be more efficient in that it requires plural judgments based on detailed comparisons of stimulus features, whereas the RH is better at tasks that require easy comparisons of dissimilar features.

Cohen (1973) proposes that the most important difference between the two sides of the brain has to do with time, so the LH is a serial processor and the RH a parallel processor. In a visual search experiment, subjects were presented with an array of two to five letters in the left- or right-half visual fields. The subject's task was to indicate whether all of the letters were the same or not. It was found that the reaction times for the same responses increased with the number of letters when the array was presented to the right visual field (RVF) but remained about constant when the array was presented to the left visual field (LVF). These results are consistent with the proposition that the LH processes information serially and the RH simultaneously, that is, in parallel. When the array consisted of unnamable shapes instead of letters, both hemispheres appeared to employ parallel processing. These results, by themselves, at first suggested that serial processing might be confined to verbal stimuli. However, other investigators were unable to replicate Cohen's results. Polich (1980, 1984), for example, found an overall LH advantage with linearly increasing reaction times for both visual fields, an effect that was found irrespective of whether verbal or nonverbal stimuli were employed. He suggests that the generally observed LH advantage for visual search could be due to the superiority of that hemisphere for fine-grained feature analysis rather than to a fundamental hemispheric serial-parallel processing dichotomy. Other studies, using a visual search paradigm, found increasing reaction times as a function of the number of items for both visual fields (Klatzky and Atkinson 1971). Studies by Young and Ellis (1985) and by Ellis, Young, and Anderson (1988) on the effect of word length on visual field asymmetries clarify the parallel-serial dichotomy. For familiar words, LVF (RH) performance declined with longer words, but RVF performance did not decline. But for unfamiliar words and nonwords, the effects were equal in the two visual hemifields. Bouma (1990: 61) suggests that these results, together with the results of numerous other studies, show that familiar words are processed serially, one letter at a time, in the RH but in parallel in the LH (see Bub and Lewine 1988), whereas unfamiliar words are processed in a serial, letter-by-letter fashion by both hemifields.

Bogen has emphasized that while the "propositional" information processing (usually) is of the LH and "appositional" processing is of the RH, it is also the case that, in minor fashion, each contains the seeds of its own other, so there is a presence of apposition information processing in the LH and a presence of propositional thought in the RH. It would seem that these experimental results, considered as a whole, also express this idea. The studies reviewed suggest that the RH is *primarily* a parallel processor but also has a *minor* capability for serial processing; the LH, in contrast, is primarily a serial processor but also has a capability for parallel processing. Similarly, Bouma (1990: 68) reviews experimental data suggesting that while there is an overall specialization, the RH can process visual information analytically, and the LH can process information holistically, although not with equal competence.

The RH is specialized for spatial, configurational, gestalt-oriented processes, such as face recognition, visual orientation, topographical orientation, and the tonal aspects of music (Bogen 1969), in which information is processed simultaneously. The RH carries out a kind of patterned thinking. Bergland makes this clear: "Your right brain ... thinks in patterns, or pictures, composed of 'whole things', and does not comprehend reductions, either of numbers, letters, or words" (1985: 1). The RH's cognitive processes are not based on temporal sequence but rather function in accordance with a principle of *symmetry*. The LH, in contrast, is specialized for verbal and nonverbal information that is *asymmetrical, that is, processed as a sequence.* The LH, for example, appears better equipped than the RH for making the fine temporal discriminations that are required for decoding a sequence of speech sounds. Insofar as "linear" time can be thought of as an ordered sequence of moments, it is a specialization of the left cerebral hemisphere. On this, we have referred to Bogen's statement that the most important distinction between the left and right hemisphere modes of information processing is the extent to which a linear concept of time is involved in thought (1977: 141). Efron (1963) shows that temporal discriminations are performed by the LH, involving the function of the temporal lobe (and in particular, Wernicke's area and the angular gyrus). He also proposes that the conscious comparison of any two sensory stimuli requires the use of the hemisphere dominant for language functions (for dextrals, usually the LH). Carmon and Nachson (1971) showed that LH damage results in poor perception of sequence, while RH damage does not. Clarke, Assal, and de Tribolet (1993) found a RH-lesioned patient impaired in time planning: she was unable to use bidimensional timetables but relied on a linear list of assignments, this speech-related strategy sustained uniquely by her LH.

Ordinary-linear time is conceptualized as a line, which by definition is equivalent to the ordered set of all real numbers. Some people with vivid mental imagery are able to visualize numbers arranged along a line (Galton 1880). Moyer and Landauer (1967) bravely claim that *all* people have a mental number line of sorts, which they estimate is visualized in about 5 percent of cases. They measured the time it took for their subjects to choose the larger of two single digits by flipping either a right-handed or left-handed switch. They found that the smaller the difference between the two numbers, the longer it took: deciding between six and seven, for example, took a tenth of a second longer than deciding between one and nine. This "distance effect" suggests that the brain was converting the digits into analog magnitudes, such as line segments, before comparing them. The proposed "number line" has been shown to be localized to the inferior parietal cortex of the LH; this brain region carries out a variety of functions, as can be inferred from studies showing that inferior left parietal lesions not only disorganize the number line (Dehaene, Dehaene-Lambertz and Cohen 1998) but result in acalculia, confusion of left and right, difficulty in finger naming, finger agnosia, and agraphia.

Dehaene has speculated that there might be a literal number line hardwired into our brains, so each number corresponding to a dedicated cluster of neurons, with the clusters arranged one after another in the same order as the numbers themselves. This, if true, would be a simple mechanism, and we do know that vision works in this way: an image falling onto the retina is mapped on a point-for-point base onto neurons in the visual cortex without disrupting the image's geometry (Kunzig 1997). Number, then, might be one of the fundamental dimensions along which the brain constructs the world. Thus it would seem that ordinary-linear time consciousness, which sees time as a numbered line, is one of the fundamental dimensions along which the brain constructs the world. Only the LH can do such a calculation. This is shown, indirectly, by research with patients who have undergone corpus callosotomy (or cerebral commissurotomy), the famous "split-brain" surgery of last resort, carried out to relieve the frequency and seriousness of severe, even grand mal, drug-refractory, epileptic seizures. When arithmetic problems were briefly flashed to the right-visual half fields of these patients, in which case the information was made available only to their LHs, they were able to solve elementary arithmetic problems. But when these images were shown to the left of these patients' midline, the RH was unable to solve such problems. Research by Wynn (1996), and replicated by Dehaene (1997; also see Dehaene, Dehaene-Lambertz, and Cohen 1998), shows that babies are born with a number line that allows them to grasp small quantities, up to three and perhaps four, and even perform elementary arithmetic on them, being able to count motions as well as objects. The number line carries out only crude calculations. For arithmetic problems beyond the working of the number line, numbers have to be represented in the brain not only as digits but also as words, because the elementary facts of arithmetic would appear to be placed in the memory in words. These words, including verbal sequences that have been learned by rote, such as the multiplication table, would appear to be processed near the center of the LH, in a multipart structure, the basal ganglia. Brain imaging of normal people, using PET and MRI scanning, shows that the inferior parietal cortex is active during most number processing, and the left basal ganglia are active during multiplication (Kunzig 1997).

THE DEVELOPMENT OF SPATIAL AND TEMPORAL ORDERING

Studies of preference for spatial and temporal ordering in recall of visual information have largely been carried out with children of urban, Western origin (Altom and Weil 1977; O'Connor and Hermelin 1973a, 1973b). In one such study, Freeman (1975) found significant evidence of a linear progression from the recall of visual information by spatial position to recall of this information by temporal position. He inferred that as "normal" children age, they come to respond consistently to temporal patterning of stimuli but not to spatial patterns. Insofar as the left and right sides of the brain are involved in spatial and linear-temporal information processing, respectively, we see a process of development first of RH-dependent spatial reasoning and second of LH-dependent temporal reasoning.

Thus there is a developmental shift of cerebral lateralization, from an early right hemisphericity to a later left hemisphericity, a process that is much in evidence by age eight and can be seen to culminate, in adolescence, in the attainment of a mode of cognition described by Piaget ([1952]1963) as "formal operations." Bogen and others (1972) define "hemisphericity" as a tendency to rely primarily on the resources of one hemisphere or the other. This process of cognitive development, as Freeman (1975) acknowledges, has exceptions, including deaf and mentally retarded children, who persistently show a recall preference for spatial as opposed to temporal ordering (O'Connor and Hermelin 1973a, 1973b). Freeman (1975) suggests that the exceptions to the spatial-to-temporal pattern result from a slower rate of acquisition of the encoding skills necessary for temporal ordering.

Davidson and Klich (1980) have found continuing preference for spatial ordering with increasing age in Australian Aboriginal children, which they attribute not to slow acquisition for temporal ordering or to any developmental abnormality but rather to "a deliberate adaptive response to cultural and ecological demands which are beyond those experiences afforded that child by a middle-class home life or schooling system in a Western urban setting" (569). Davidson (1979) reviewed ethnographic descriptions of cognition in traditional Aboriginal societies—including card playing, orienting, and ceremonial instruction—and concluded that these behaviors involved primarily the synchronous or simultaneous rather than the serial or successive synthesis of visual and other information. He suggests further that Aboriginal children are likely to develop simultaneous-spatial skills rather than sequential-temporal skills, with their trend toward temporal ordering increasing only minimally with age.

The gestalt orientation of Aborigines might well have originally resulted from the cognitive demands of life in a hunting-and-gathering society. But this mode of thinking is likely to persist even in an urban environment, where Aborigines are apt to live a marginal, fringe existence. Under almost all circumstances, Aborigines have held tightly to their extended families, and this continuing experience of communal-sharing social relationships, it is proposed, contributes to a gestalt-synthetic mode of thought. For example, in the fringes of Darwin, Sansom contends, there exists for the Aborigines "a total state of affairs today, another totalization tomorrow" (1980: 190). This gestalt summarization is not a function of the Aborigines' putatively artistic nature but is rather borne of necessity. As Sansom puts it, this particularization into existence of the fringe community "is to supply the sort of information that allows for a running accounting when, for social survival, only running accounts will do" (1980: 191). The mode of thought of Aborigines, both children and adults, emphasizes a gestalt summarization that is constantly updated (Grey 1975).

Time is fundamental to an understanding of the real world. Time is not simple but composite. Time is a hierarchy of more and more complex temporalities. Temporally sequenced events are ultimately analyzed by the logical-analytic information-processing capability of the left side of the brain, spatially simultaneous events by the gestalt-synthetic information-processing capability of the right side of the brain. No events, and no psychological stimuli, are either purely temporal or purely spatial. A mode of information processing that is primarily temporal and secondarily spatial is defined here as "spatiotemporal," the primarily spatial and secondarily temporal as "temporospatial." Thus it is proposed that the right and left hemispheres of the human brain are specialized for levels of information processing that are predominantly, but not exclusively, *temporospatial* and *spatiotemporal*, respectively. This formulation is consistent with Levy, who has argued that "the left brain maps spatial information onto a temporal order, while the right brain maps temporal information onto a spatial order" (cited in F. Turner 1985: 70). The two "brains" transfer information to and fro by the corpus callosum, alternating in the "treatment of information according to a rhythm determined by the entire brain state and pass, each time, their accumulated findings on to each other" (F. Turner 1985: 70), which very well might take place *under the leadership of the frontal lobes*, acting as an "integrating agent between left and right [hemisphere] functions" (Gunther Baumgärtner, cited in F. Turner 1985: 71).

Chapter 7

Immediate-Participatory and Episodic-Futural Time and the Brain

Now that we have developed seven-part definitions of patterned-cyclical and ordinary-linear time consciousness and established their infrastructure in the modes of thought of the left and right sides of the brain, we can turn to a second temporal distinction, that between *immediate-participatory* and *episodic-futural* kinds of time consciousness. The two kinds of temporal experience will be described, and in an effort to establish criterion validity, their possible biological infrastructures will be explored and their limitations discussed.

IMMEDIATE-PARTICIPATORY TIME CONSCIOUSNESS: MINDFUL EXERTION IN THE PRESENT MOMENT

William James (1890) and Albert North Whitehead (1929) developed complementary philosophies of the unifying moment that have been clarified and elaborated on by Eisendrath (1971). Both saw ideas as being constructed out of our immediate experience of the sensed world. James (1890) wrote of the organization of the conscious field and of its development in constructive mental work that takes place at every moment, this construction of the moment being based on the fact that materials separated in clock time can be related in consciousness. He provided an example of this knowing of things together: "Into the awareness of the thunder itself the awareness of the previous silence creeps and continues; for what we hear when the thunder crashes is not thunder pure, but thunder-breaking-upon-silence-and-contrasting-with-it" (1890, 1: 240). Eisendrath (1971) explains: "For a complete relation to be felt, its relata, the experience of the 'past' and the 'present' object must be co-present psychic experiences" (48). James insisted that awareness of *change* is a fundamental condition on which our perception of time's flow depends. The explanation offered for this awareness of change, by James (1890, 1, 608) and Whitehead (1929), is that of the "specious present." In this immediate, participatory temporality, the world is perceived, and understood, as "a continuous moving picture" (Hodgson 1991: 13). This reality, in our conscious states of mind, synthesizes the perceived actuality of the world as it is and is becoming. This level of consciousness must be considered to last for some duration of time, this duration defining the *thickness* of the present, the conscious experience of the moment, as it is difficult indeed to make sense of consciousness unless it is considered as lasting for some time. And consciousness is often a *succession* of events, which we perceive as a continuing scene in which there is both movement and change.

Of course, touch, smell, and "feel" are important to perception and to the experience of time, but it is primarily sight and sound that occupy Hodgson's "slab of time." The infrastructure for this mental activity is, in Hodgson's terms, "being constructed by a complex web of events, occurring over regions of the brain and communicating with each other ... within the finite timesspan of the psychological present" (1991: 13). For vision, the seen object can have characteristics that endure, such as color and shape. Hodgson (1991: 13) provides the example of a red circle, where the redness and the circleness are inextricably together, not successive. Yet different groups of neurons are involved in circle detection and in red detection, so there must be a form of nonlocality of thought to explain the "all at once" awareness of the red circle. Perception of objects requires both largely unconscious, hierarchically structured, information processing and ideation of the whole moment.

The brain does not work like a closed physical system or a computer but rather as a system of nonlinear oscillators. The bracketing of objects and the chunking of information in short-term memory bring together elements associated with events in spatially extended but largely posterior areas of the brain. The scene before us involves shapes, colors, and movement, the consciousness of which "knowing together ... a totality of elements corresponding to many spread-out events" (Hodgson 1991: 6). One such minimum duration, about 20 milliseconds, is enough time for millions of neural information transmissions to take place and "for all kinds of communications to take place between ... parts of a human brain" (Hodgson 1996: 12). The representation of this web of interrelated event-processes in the brain requires the participation of a wide variety of spatially separated brain events involved in forming a perception of the moment.

In immediate, participatory time consciousness, there is a *temporal compression* of the past and of the future into the present. This important concept can be better understood by considering the thought of thirteenth-century Japanese Sōtō Zen philosopher Dōgen Kigen (1972; lived 1200–1253), who formulated a temporality of the immediate moment in part as a critique of linear time. He called this temporality "being-time" (in Japanese, *uji, u* meaning "exist" and *ji* meaning "time"), a temporality existing entirely in the present Moment, meaning there is an immediate and a complete realization in the Now, centripetal, a "taking place" (*kyōryaku*) during which the past and the future are *compressed* into the present. Ontological truth, the accomplishment of being-time, results from "total exertion" (*gūjin*) and is spontaneous, realized in the here and now, and realized with great difficulty and real exertion but without hesitation or expectation (Kim 1975: 206). Dōgen saw the truth of time (*dōri*) as the basis of the utmost exertion of the skin, flesh, bones, and marrow of the one who has penetrated the meaning of, and who has accepted, finitude. Such exertion can be authentic, he realized, in what he described as his moment of enlightenment, only to the extent that it affirms the *impermanence of all things*, the unsubstantial and dynamic nature of *uji* (Heine 1985: 125). Immediate-participatory temporality encompasses human and natural sameness and difference without dichotomization and prior to the categories of the understanding.

This level of temporality is insubstantial insofar as social and other activities are carried out without reference to subject or object, the mind's orientation rather being the ebb and flow of movement itself. The result of this exertion is the multichannel and holistic presencing of naturalistic activity, an elementary form of thought that is of the moment, immediate, and fully involved in the sensed, natural world. The lived body in this being-time is not an entity set apart from but rather engaged in the world. The present moment thus has a completeness that is independent of anticipation or expectation, because it is not going anywhere or coming from anywhere. Every Moment, then, is an impermanent and insubstantial unity. Past, present, and future are but provisional terms for conveying the totality of dimensions that are present at *every* moment (Heine 1985: 130), which requires the utmost exertion in the moment of an active *participation* in the world in which "Past and future continually merge in the present" (Stambaugh 1990: 37–38). This primordial temporality is thus both immediate, compressed into the Now, and participatory. The sensed world is not static but rather filled with change and activity. Because no thing, and no being, is permanent, the existence of a thing, its "to be," means to take place, to make a passage (*kyōryaku*) that is *right now* (*nikon*) and that "encompasses the simultaneity of past, present, [and] future" (Heine 1985: 153).

This participatory temporality is not to be found in a passive-dependent, relaxed, drifting immersion in the present. Of course, every individual person and members of every culture, to some extent, are continuously oriented to the present, to the sensed world that is ever here and now. Only in deep sleep and in other forms of unconsciousness are there time-outs from the processes of sensing and perceiving. This temporal orientation has been described by Kluckhohn and Strodbeck (1961) as a "timeless present" and by Langer (1997) as a state of "mindlessness," which has the consequences of accidents, burnout, poor job performance, memory problems, interpersonal difficulties, and health problems. Mindlessness turns the body off, blocking our mental awareness of what is going on around us. When mindless, we let the past determine the present, and we do not pay close attention to what is going on around us. In a mindless state, we become unaware of the small, unexpected changes and differences that can warn us of possible difficulties; we become insensitive to context and to perspective. It is when we are rather "mindful" that we notice new things, noticing differences in things we think are similar, and noticing similarities in things that we consider different. Immediate-participatory temporality, then, demands mindfulness. It is brought about, as Dōgen insisted, through real exertion in the now, in the knife-edged present moment.

We have seen that patterned-cyclical and ordinary-linear time consciousnesses are opposites, made so by a definitional process, clarified and elaborated on by a review of their biological infrastructure, which suggested that they process information *sequentially* and *simultaneously*, respectively. We will now see that immediate-participatory and episodic-futural temporalities also are *opposites*, in the sense that immediate-participatory time consciousness requires a temporal *compression* and episodic-futural time, as explicated by Heidegger ([1927] 1962: 423), and requires a temporal *stretching*. There is a *complementarity*, however, between immediate-participatory and patterned-cyclical time consciousness, however, for both take place, simultaneously, in a moment of perception or a moment of closure. This topic will be pursued in chapter 9.

Episodic-Futural Time Consciousness

Heidegger explicated a kind of primordial temporality that moves centrifugally, *stretching* to the horizon of our very Being. The human being experiencing this time consciousness stands, ecstatically, within the opening of being, holding and keeping it free and open. In his *magnum opus, Sien und Zeit* (*Being and Time*), Heidegger ([1927] 1962) analyzed human existence on the level of everyday life. From this vantage point, he was able to make topical the Being of *Dasein*, of human existence, and was then led to temporality, which unifies *Dasein's* Being. He used the term *time* when referring to objects in the world, to the world itself, and to relations between the subject-knower and object-known. In his effort to transcend the subject-object distinction, his "temporality" refers not to subject and object but rather to *Dasein's* existence, which, as temporal, "creates" time and "sees" time, so that time is grounded in temporality and in the hard fact of finitude. Being always possesses a "temporal determinateness" (*temporalae Bestimmtheit*). By this, Heidegger meant that Being is not an abstract, independent realm that transcends time but is rather dynamically revealed by the *horizon* of time. His primordial time (*ursprüngliche Zeit*) is an experience reserved for those people who understand what it means for things to be and who possess the will to face the certainty of a future in which being itself comes to an end. *Authenticity* is to be found, Heidegger claimed, where openness to the future means both acceptance of death, of the inevitable erosion of time, which defines temporality, and also of a world renewed and filled with possibilities (Zimmerman 1981: xx).

Insofar as we are, as Heidegger argued, fundamentally "ecstatic" in the temporal experience, in Richardson's terms, "we reach ahead to our ends, from out of a rootedness in what we have been, and through (or by means of) the entities with which we are now preoccupied" (1986: 94). Heidegger's notion of ecstasy with respect to temporality is far from clear but would appear to involve a commitment of emotional energy, of an *élan vital*, which, according to Minkowski, "creates the future before us.... In life, everything that has a direction has *élan*, pushes forward, progresses toward a future.... The *élan vital* discloses the existence of the future to us, gives it a meaning, opens it, creates it, before us" ([1933] 1970: 38). The future, as constituted in the mind, is therefore based both on emotion and reason as intentions are turned into *acts*, for episodes of very brief duration, and for *actions*, for episodes of longer duration, stretching out into the cared-about future in an effort to attain objectives, carry out plans, and realize intentions.

The past of our temporality cannot be reduced to a sequence of nows left irrevocably behind but is rather the understanding we have of what we have been and that out of which our current projects and actions arise. Every person's activities are guided by his or her attunement to the past. We are thrown into, and enveloped by, moods, and we experience emotions by reaching back into our memory of the past and encountering present things from out of that past. Nor can the future of our temporality be reduced to a span of moments not yet manifested. The future is rather the "horizon" toward which we are projecting, perhaps as a familiar pathway on which we are en route or to a laudable goal that we are seeking with determination, confidence, and resolute intentionality. It is only because of this horizon and looking to the future that our present experience of entities is possible. In our everyday and mundane world, we ordinarily lack clarity and explicitness in our imagination of the future and our efforts to accomplish some future state of affairs. In fact, we only sporadically reflect upon our ends, and we lack full awareness of these ends. Our actions are necessarily based on imperfect judgments and anticipations which, in Dewey's terms, "can never attain more than a precarious probability" (1929: 6; see also Dewey 1925: 41) as we head toward an aleatory world that we can see only as a scene of risk and instability. Whenever we undertake a task or project, our temporality is stretched out all along the course of, and in some circumstances beyond, the project, which requires an implicit familiarity with what came before and what will come after, and with how present activities fit within an overall effort (Richardson 1986: 107). Thus Heidegger's notion of primordial temporality is based on an anticipatory resoluteness and is futural and ahead of itself insofar as we not only contemplate the realization of objectives and aims but also imagine the consequences that will follow from success or failure in such striving.

Heidegger has identified an elementary, irreducible form of temporality. This temporality, determined primarily out of the future, is stretched over a period of linear, calendrical time, as we will see in chapter 9. The phenomenal experience of that which is stretched will be referred to here as an "episode." The reality of primordial temporality is to be found on the level of thought. We have seen in Heidegger that anticipatory resoluteness is the essence of effort to bring into being an *anticipated*, and, on the level of emotions and desires, a *cared about* future state of affairs. Several related mental actions characterize the mind's effort to attain an anticipated future state or end—such as intending, planning, managing, anticipating, monitoring, editing, commanding, controlling, and more generally carrying out the executive functions of the mind.

Immediate-Participatory and Episodic-Futural Time Consciousness and the Brain

Pribram (1981) identifies the "participatory" and the "episodic" modes of information processing that he attributes (approximately) to the functioning of posterior and frontolimbic areas of the brain, respectively. It is proposed that the immediate-participatory and episodic-futural kinds of time consciousness are expressions of the *primary* sense perception of the posterior cortex and the episodic conation of the frontal lobes, respectively. Major corticocortical connections between the prefrontal and sensorial areas of the brain will be described, and it will be shown that the interaction made possible by these connections is essential to the performance of the highest mental functions, to the modeling of the world, to making plans, and to acting with intentionality.

Channels of sensory information (other than olfactory) enter the cortex at the primary association areas behind the frontal lobes. In both the left and right sides of the brain, the parietal, occipital, and temporal sensory areas send information to adjacent "secondary" and "tertiary" association areas in these same posterior lobes (Fuster 1980), where there is a lateral division of labor, also sent to the prefrontal cortex. The prefrontal cortex is thus strongly connected to the modal and multimodal sensory association cortex—for the important senses of sound, sight, feel, and taste—and functions as a supramodal control center able to anticipate novelty. The prefrontal cortex receives less information from the primary sensory and motor areas than it does from the secondary and tertiary association areas, largely located posterior to the frontal lobes. This is highly significant, as it implies that much of the information transmitted to the prefrontal lobes is not raw data, rather having been subjected to secondary and/or tertiary information processing. Thus much of the sensory information received by the prefrontal cortex has already been subject to cognitive processing (e.g., by the logical-analytic and gestalt-synthetic information processing lateralized to the LH and RH). Thus there exists throughout the *waking, conscious* state of mind a never-ending complex, multilevel "conversation" between the prefrontal and sensorial areas of the brain. Major pathways originating in the somatic, visual, and auditory systems converge on contiguous but discreet areas of the prefrontal cortex (Fuster 1980). Among its many other duties, the prefrontal cortex functions as a multimodal sensory association area, concerned with "egocentric spatial orientation" toward events in sensorial space, and it persistently integrates information about these events. Sensorial space itself is constituted, in large measure, by association networks in the *parietal* cortex. Motor activity within this sensorial space also is integrated into these parietal association networks (Laughlin 1988: 249).

While episodic processing reestablishes stability through "chunking" experience into episodes for higher-level processing (Pribram and Tubbs 1967), at the cost of simplification, participatory processes tolerate transience for the gain of flexibility. "Participatory" processing deals with incongruity by searching *and sampling the input and accommodating the system to this input*. Equilibrium is not a reestablishment of an earlier status quo; plans of actions can rather be adjusted and even changed to fit a new reality. Our model of the world, when found inconsistent with data from the senses, is restructured, so that it can, hopefully, again function in a stable way.

Pribram (1981) proposes that "episodic" reasoning is the mode of information processing of the frontal lobes of the human brain, working in conjunction with the limbic system, posterior cortex, the reticular activating system, and nearly every other area of the brain. The frontal lobes of the brain are essentially involved in carrying out *reflective thought, reason*, and *conation*, in which attention is not immersed in the moment but is rather stretched out across an episode. An "episodic process" prepares the organism for further interaction with the world by encoding the structure of redundancy as the context within which subsequent action is framed. "Planning" is part and parcel of episodic reasoning; the structural basis for planning, when broadly defined, is the frontal lobes (Luria 1966, 1973). The frontal lobes abstract certain features from perceptual images and recombine these abstractions into models, which form the basis of decision making and action. Sensory inputs not screened out by habituation and gating mechanisms are fitted into these images, or used as indexical summaries in episodic processing. These abstract mental images, or category prototypes, of prospective conduct enable the rehearsal, and then the actual carrying out, of acts, actions, and activities.

The frontal lobes regulate the "active state" of the organism, control the basic elements of the subjects' intentions, program complex forms of activity, and continually monitor all aspects of activity (Luria 1973: 187–225). In order to act with intentionality, the frontal lobes must be able to evaluate the results of one's own actions. They carry out a complex process of matching actions carried out with initial intentions, to evaluate success and error, so that actions can be monitored, then corrected, edited, and modified as necessary. The frontal lobes thus constitute the command-and-control center of the brain.

The prefrontal cortex is the latest development in the evolution of the human brain. In the development of the individual person, it is the last area of the cortex to myelinate, is phylogenetically the most recently evolved cortical area, and has exhibited allometrical greater development than most other areas in hominid brain evolution (Passingham 1973). The frontal lobes are arguably the highest achievement of human brain evolution (Halstead 1947). They are strongly connected to portions of the mediodorsal nucleus, phylogenetically the most recent thalamic area to evolve and the highest association area in the brain.

While frontal-lobe lesion patients are comparable to normal patients on many intellectual tasks—which after all can be solved through analysis and synthesis, the most global lateralized functions of the posterior cortex—Milner (1964) and many other neuroscientists have found that frontal-lobe patients with dorsolateral (but not orbital) lesions are unable to effectively shift their responses to meet changing environmental goals. These patients do not suppress their ongoing response tendencies, whether these tendencies were spontaneously or experimentally induced. Deficits in the usual metalinguistic monitoring of speech by the frontal lobes can be seen following lesions of the left frontal lobes (the inferior lateral-frontal regions), as these patients lack verbal fluency, produce less spontaneous speech, and do not correct errors in their speech and writing. The frontal lobes play a modulatory function, which can help explain the inadequate social behavior and coexisting good performance on many standardized intelligence tests following frontal-lobe damage. Luria (1969), among many others, has observed that damage to the frontal lobes disturbs impulse control, the regulation of voluntary actions, and perceptual processes such as visual search. Such damage adversely affects memory components that are involved in strategic planning and in symbolic functions, because it results in poor choices of programs to carry out and a lack of ability to restrain premature operations (Das 1984: 36). With damage to the prefrontal cortex, there is a loss of will, a flatness of affect, and a loss of planning and anticipation. Frontal-lobe patients show the lack of an active, future-oriented attitude to the world. It is through the prefrontal cortex that we are capable of sustained concentration on plans and programs and are able to act in the world with an appropriate level of confidence in our plans and intentions.

The brain regions most involved in intentional functions are the dorsolateral and orbital prefrontal cortex (Pribram 1971; Pribram and Luria 1973; Stuss and Benson 1986; Fuster 1980). The principal and lateral dorsal limbic nuclei have an absolutely, and a relatively, greater number of nerve cells in modern humans than in studied species of great apes (*pongids*) or lesser apes (*hylobatids*). The larger size of these features might well modulate the integration of emotion and cognition, relaying a larger emotional component to the posterior cingulate gyrus (attention) and posterior association areas (Armstrong 1982, 1991). Episodic processing, then, is not a neutral instrument of rationality but rather is closely bound to emotional responses, which are important because the exercise of human reason requires a close interaction between "rational and emotional proclivities" (Boyle 1985: 65; de Sousa 1987: 171–204 passim; Barbalet 1998: 29–61).

Thus an "episodic" process prepares the person for further interaction with the environment. Environmental input is selectively structured according to its relevance and usefulness to the ongoing plan of action. Irrelevant input is screened out, enhancing redundancy and conserving former plans. Relevant input is taken in, leading to complexity and uncertainty, which calls for modification of plans. "The achievement of external control is conceived through the accommodation of past experience to current input to lead to what is subjectively felt as satisfaction" (Pribram 1981: 121). A sense of satisfaction results from similarities that are identified between past experience and concurrent input. The achievement of internal control comes about through the fulfillment of intentions, or the restoration of ongoing plans, which results in the subjective experience of gratification as things turn out as intended.

Laughlin (1988) has developed a conceptualization of relations between the front and back of the brain, and the associated mental processes that importantly bring into view the notions of subject-object distinction, intentionality, and the self. These dynamic relationships, the *prefrontosensorial polarity principle*, link the prefrontal and sensorial areas of the brain, mediate intentional processes in consciousness, and contribute to the sense of a subject-object distinction. While Heidegger sought, unsuccessfully, to transcend the subject-object polarity on the level of his philosophy of Being and time, the neuroscientific evidence suggests that this distinction is embedded in the very structure of brain organization and is ineradicable, being of practical value in the everyday world and usually a taken for granted capability of the human brain. According to Laughlin, the most fundamental intentional functions are mediated by the prefrontal cortex, all of which have to do with objects: (1) "the anticipation of, selection of, orientation toward, concentration upon, and cognitive operations upon the phenomenal object abstracted from its sensorial context"; (2) "the systematic inhibition of irrelevant sensorial objects and events, as well as affective and other neural activities competitive with the object of the intentional process"; and (3) "the establishment of a point of view relative to sensory events, and under certain conditions of a cognized distinction between self and other, or subject and object" (1988: 251–52).

Boyle (1985) addresses the concept of "balance" between episodic and participatory information processing, asking which kind of balance is implied by Pribram's model. First, it might be the case that one mode makes a greater contribution to awareness than does the other. This suggestion is consistent with Heidegger, who saw in the modern individual a future-value orientation, which we can now call a "primacy of episodic conation over participatory processing."

The theoretical continuity between the work of Pribram (1971, 1981) and Luria (1973, 1982) suggests a possible conceptual synthesis, which has been successfully carried out by Boyle, in a way that brings the concept of self into our understanding of episodic and immediate forms of temporality. Boyle (1985) argues that mental dialogue contributes to episodic processing by constructing, editing, and revising internal models, with three consequences: (1) Insofar as inner speech does the actual modeling work, the left-frontal cortex contributes to episodic processing; (2) There is an emotional substrate of such episodic models that is associated with the workings of the right-frontal cortex, thus linking the right hemisphere to mental dialogue. And (3) the potentials of the left frontal cortex are not fully exhausted by episodic processing. Ideas are somehow "generated" through processes involving the frontolimbic area of the brain. Ideas are to be distinguished from episodic processing, because ideas (but not episodes) are synchronically rather than diachronically organized. Ideas are able to contribute to episodic processing through being translated into the symbolic, linguistic organization of the episodic process, but they are encoded more like the perceptual images of participatory processing and the visuospatial gestalts of the right hemisphere.

We have seen that executive processes are responsible for the volitional control of cognition and for the regulation of thought and behavior. The study of executive processes has developed together with the notion of frontal-lobe involvement in such processing. However, serious problems remain. Tests of executive functions have low test-retest reliability and uncertain validity. It has been difficult to clearly distinguish between "executive" and "nonexecutive" tasks. Behaviors that would appear paradigmatic exemplars of executive behavior have proven difficult to link to specific neuroanatomical areas of the brain (see Rabbit 1997: 1–35 passim; Zelazo et al. 2003). What would appear as "components" of executive functioning, such as planning, monitoring, and controlling, typically have poor construct validity and are in fact "simply descriptions of task demands" that appear "logically different" but that "can be met by identical production system architectures" (Rabbit 1997: 1). Most cognitive neuropsychologists find it difficult indeed to give a satisfactory functional account of the "central executive" system, and some philosophers (e.g., Fodor 1983) declare it impossible. Rabbit (1997: 2) points out that executive impairments associated with frontal-lobe lesions are "strikingly similar to definitions of 'willed,' 'purposeful,' or 'voluntary' behavior that have preoccupied philosophers and theologians for more than two millennia. For example, contemporary catalogues of the functions of the hypothetical 'central executive' are strikingly similar to the formal criteria for commission of mortal sin given by Roman Catholic theologians" (Rabbit 1997: 2). At the same time, there remains little doubt that executive control is, again in Rabbit's terms, "necessary to deal with novel tasks that require us to formulate a goal, to plan, and to choose between alternative sequences of behavior to reach this goal, to compare these plans in respect of their relative probability of success..., to initiate the plan selected, and to carry it through... until it is successful or until impending failure is recognized" (1997: 3). Given this statement by a scholar who presents criticisms of effort to *measure* central executive function, we can see that, such difficulties aside, there remains an unshakable linkage between the functioning of the frontal lobes and episodic, future-oriented conation.

Chapter 8

The Two and the Four, and Possibly More
Social Duality and the Four Elementary Forms of Sociality

Now that the first, cognitive model of the present theory, specifying the existence of four elementary forms of temporality, has been presented and its biological infrastructure has been examined, we turn to the other model, that of social relations and, more generally, of societal organization. This model is derived from conceptual continuities in primate ethology, psychology, neurobiology, and sociology. It proposes that there exist four (and of course, possibly more) elementary problems of life—identity, reproduction and the life cycle, hierarchy, and territory. In human society, these problems of life have come to be addressed by four elementary social relationships—equality matching (EM), communal sharing (CS), authority ranking (AR) and market pricing (MP) (Fiske 1991, 1992). Together, EM and CS engender *hedonic* society, AR and MP *agonic* society. Hedonic and agonic kinds of society are interpreted as opposites on the grounds that their component elements are opposites, with equality matching and authority ranking being opposite in meaning and communal sharing and market pricing being opposite as well.

Each of the four problems of life then will be paired to one of the four elementary forms of temporal experience. Chapters 10–13 will link patterned-cyclical time to CS, immediate-participatory time to EM, episodic-futural time to AR, and ordinary-linear time to MP. The time consciousnesses of Aborigines and Westerners corresponding to the four kinds of social relations also will be considered. The resultant theory of time and social relations will then be given empirical content through a comparative study of Australian Aborigines and Euro-Australians.

Social Duality Theory, Unpacked

We begin with a consideration of what can be called "social-duality theory." The idea that there exist two kinds of society has ancient roots and finds modern expression in social science and primate ethology. Such dualities have been proposed by Confucius (the Small Tranquility, the Great Similarity), Plato (the Ideal Republic, Oligarchic Society), Aristotle (True Friendship, False Friendship), and St. Augustine (the City of God, the Society of Man [*corpus mysticum*]). Similar dichotomies have been offered by Joacham de Fiore, St. Thomas Aquinas, Nicolaus Casanus, Ibn Khaidun, Savigny, and Hegel (Family Society, Civil Society/the State). The distinction, which in all of these examples is that between formal and informal social organization, also has found expression in Tönnies' ([1887] 2000) distinction between the folk *Gemeinschaft* (community) and the exchange *Gesellschaft* (society), and in their associated mentalities of Natural Will and Rational Will, respectively, in whose honor the *natural-rational* distinction in time experience is introduced, and in that between Durkheim's ([1893] 1933) notions of mechanical and organic forms of social solidarity.

A more recent, and more useful, model of social duality can be found in primate ethology. Through comparative study of the behavior of the higher primates and of human evolution, it can be seen that Chance's (1988) distinction between agonic and hedonic forms of societal organization is fundamental to primate social organization. This agonic-hedonic distinction is valid but is at the same time theoretically underspecified, a limitation that can be resolved by conceptually "unpacking" these two concepts, to reveal that each is based on two complementary social relationships.

The Agonic Model of Societal-Level Organization

While the agonic model characterizes Old World monkeys (the primate superfamily *Cercopithecoidea*) such as macaques and baboons, the hedonic model appears in the *Hominoidea* (apes and humans). Agonic-type societies are conflictual and hierarchically organized. This model of society can be clearly seen in nonhuman primate societies in which individuals are arranged in a series of status levels. Any two individuals in such a society are thus either of the same or of different social rank. The difference in rank between individuals manifests itself in the acquisition of what is of social value, including attention. Each individual gives and receives attention according to his or her rank. Higher-ranking individuals receive more attention than they give; lower-ranking individuals, in contrast, give more than they receive.

Social dominance is the primary dimension of agonic society. Dominance involves both hierarchy and control of resources. Those higher in rank control the behavior of those lower in rank. This control is expressed territorially, through the proximity of the lower-ranking members to the centrally dominant figure(s). Emory (1988), in a study of caged monkeys at the San Diego Zoo, found that the amount of attention paid to the dominant male was directly related to the nearness of each individual to the leader. Also observed was what Chance and Jolly (1970) have called "reverted escape"—the return of an individual to the vicinity of the dominant male after withdrawing in the face of an intermediate-level threat from the leader(s). Such a behavioral pattern can be seen in Indian Macaque monkeys and African Savannah baboons. Thus *hierarchy* is inseparable from and articulated in terms of *territory*. Hierarchy and territoriality are the basis of agonic society, the grounds of its social cohesion. Dominant males, for example, in the Chance, Emory, and Payne (1977) study of long-tailed macaques, engage in unprovoked aggression, often early in the day, as a way of testing the stability of the social order. An unexpected reaction of a threatened subdominant member, or an aggressive behavior on the part of a lesser member, also can signal instability and the potential for change in the social hierarchy.

In the agonic mode, such as in the *rhesus macaque* or the Savannah baboon, individuals are kept together in a group yet are spread out, keeping their distance from one another and from the more dominant one to whom they must constantly attend. Subdominant members are always ready to react to a threat in order to avoid punishment. They do this with gestures of appeasement or submission, and by "spatial equilibration... which, arising from withdrawal following the reversion of escape, serves to prevent escalation of threat into agonistic conflict, yet with tension and arousal remaining at a high level" (Chance 1988: 6–7).

Evidence exists for a neurophysiological mechanism capable of sustaining the continuously high level of tension and arousal characteristic of the agonic mode. Gilbert (1984) reviews this evidence, referring to this neurophysiological state as one of "braking," which "implies an unabated state of arousal which does not provide any effective behavior as long as the powerful brakes [which he speculates are controlled by the hippocampus] are applied" (109–11). Chance (1988: 7) suggests that this mechanism must be closer to the center of the central nervous system than the mechanism that maintains the braced musculature observed by Whatmore and Kohli (1974). The braking state is accompanied by the fixing of attention toward more dominant individuals from whom threats are anticipated and with respect to which self-control on the part of potentially aggressive challengers is exerted. <u>Thus in the agonic mode, the individual's attention is fixed on personal security with respect to rank behavior,</u> this concern with hierarchy being expressed in personal territoriality, with distancing, and with reverted escape from the more dominant group members. This hypersociality can be extremely dysfunctional, as it comes at the expense of an unawareness of much of the natural surroundings and the dangers it presents. A study of vervet monkeys (*Cercopithecus aethiops*) in Kenya showed that these animals were highly knowledgeable about social dynamics within their own social groups yet paid little attention to the natural world. Faced with an annual mortality rate of about 65 percent (Isbell 1990), mostly from leopard predation, and equipped with a specific alarm call for leopards, they nonetheless systematically failed to associate carcasses cashed in trees with leopards' presence, and also showed a potentially fatal inability to associate python tracks with pythons' presence (Cheney and Seyfarth 1990).

The Hedonic Model of Community Organization

Hedonic society appears in apes and humans. Power (1986) shows that the chimpanzee possesses a hedonic-type society in which members are not under threat of punishment. This can be seen as the group splits up into twos and threes to go foraging for food, when the less confident individuals seek and are offered reassurance by contact gestures such as touching and kissing, usually from older, more confident leaders. After foraging in small groups, chimps recongregate in response to "calls" that food for all has been found, whereupon "carnivals" are held that focus attention on the most dominant males, who are apt to demonstrate by jumping up and down and throwing things. Such gestures reduce social tensions that might have built up, meaning, "… except during moments of excitement *the arousal level of the individual is low—this is the hedonic condition*" (Chance 1988: 7; emphasis in original).

This freedom from social preoccupation also characterizes the human primate, so that, according to Chance, "The healthy human individual has a flexibility of arousal and attention that allows time for integration of reality, inter-personal relations, and private feelings and thoughts, providing prerequisites for the operation of a systems-forming facility" (1988: 8). Ethological studies of children in playgrounds carried out by Montagner and others (1988) led to the observation of hedonic, leader-type children who, like wild chimps, do not escalate threats into aggression and who actively appease their followers rather initiating play and other cooperative activities.

Hedonic society is based on two fundamental social relations: *temporality*, which requires no explanation, because the mother-offspring unit is center to the community, and *conditional equality*, which does require explanation. In agonic society, there is an inequality principle, with coexistence based largely on the self-restraint of the subordinate and, of course, also on the self-restraint of the dominant. A state of inequality *precedes* the state of conditional equality, the negation or suspension of inequality, in the process of primate social evolution. Conditional equality can be seen, for example, in play, where rank order is ignored or set aside and there is self-handicapping by the stronger participant (Itani 1984). Agreement also is needed to open up a fictitious world, an agreement to render inequality nonexistent. It is necessary to communicate in order to play, so "play participants attempt to form media [of communication] even out of their daily behavior by changing its ordinary tempo or rhythm" (Itani 1988: 147). The fictitious world of conditional equality extends beyond play, as it extends, for instance, to social grooming and to allo-mothering, formed on an agreement in accordance with each separate context, which must always exist in the moment, as immediate, as participatory. Itani makes a compelling argument for an astounding claim, that "the egalitarianism seen among the hunter-gatherer and nature-dependent people of today is nothing but a product of the evolutionary elaboration of its counterpart found among the chimpanzee" (1988: 148). The *demand* for equality among the group, Itani contends, "permeates every sector of life" (*ibid.*).

CONTINUITIES IN PLUTCHIK'S PSYCHOEVOLUTIONARY CLASSIFICATION AND MACLEAN'S TRIUNE BRAIN THEORY

Plutchik ([1962] 1991) bases his well-known psychoevolutionary classification of emotions on the proposition that members of *all* species of animals confront the same four fundamental, existential problems of life—identity, temporality (reproduction), hierarchy, and territoriality—which are exactly the same four principles that have just resulted from unpacking Chance's hedonic-agonic social duality. The negative and positive experiences of these life problems, Plutchik holds, lead to prototypical adaptive reactions, comprising eight primary emotions, which in turn pair to form secondary emotions. Plutchik's claim that his four problems of life apply to *all* animals is dubious. There is evidence, however, for a more modest but still important claim, namely, that the *higher* animals—reptiles, birds, and mammals—share these four problems.

This evidence can be inferred from MacLean's (1973, 1977) triune brain theory, which holds that the reptilian brain (R-complex) was the infrastructure for the evolution of the paleomammalian brain of mammals, which in turn was the infrastructure for the evolution of the two neo-mammalian brain structures of higher primates, the left and right hemispheres of the brain. In the human, the R-complex persists, roughly, as the brain stem. This theory, especially in its claim that the limbic "system" (including the amygdala, hippocampus, hypothalamus, and septum) is the seat of the emotions, has its limitations and its harsh critics (e.g., LeDoux 1996: 85–103), but here interest is limited to a noncontroversial component of MacLean's (1964, 1977: 211–12) model, which is his description of the R-complex having just four concerns—*identity, reproduction, hierarchy,* and *territory*. Because mammals and primates retain the R-complex as their first stage of brain evolution, it follows that mammals and primates, including the human, share an objective need to address these four elementary problems of life. Plutchik and MacLean apparently developed their biological models of the fundamental problems of life independently, but their conceptualizations are identical. According to Plutchik ([1962] 1991), there exist exactly four fundamental, existential problems of life, which are *identity, temporality, hierarchy,* and *territoriality*. There is a minor terminological difference: Plutchik refers to existential problem of "temporality," meaning the cycle of life, which has the positive function of "reproduction" and the negative function of "reintegration" of the group following a death or other loss of a community member, whereas MacLean, ignoring valence, simply refers to the life problem of "reproduction." The basic design of Plutchik's model is shown in table 8.1.

Temporality refers to the finite life span of all creatures, to the inescapable reality of death, which creates the inevitability of separation and loss. Individuals not supported by the group do not survive long. Distress signals following loss and death, which are widespread in the animal kingdom, are cries for social support, for sympathy, and for nurturing, which function to reintegrate the group minus its lost member. This experience is

TABLE 8.1 The four existential problems of life, and the four primary emotions processes, valences, and arithmetic operations associated with each problem

Emotional Problems	Emotional Functions	Subjective Terms	Behavioral Processes	Valences	Arithmetic Operations
Territory	exploration	anticipation	open boundary	positive	*
	orientation	surprise	close boundary	negative	÷
Hierarchy	destruction	anger	move toward	positive	>
	protection	fear	move away from	negative	<
Temporality	reproduction	joy	gain	positive	+
	reintegration	sadness	loss	negative	−
Identity	incorporation	acceptance	take in	positive	=
	rejection	disgust	expel	negative	≠

Source: TenHouten 1999a, table 1, p. 61.

definitive of *sadness* and *grief*. *Joy*, or *happiness*, is the opposite of sadness, which is, of course, associated with sexual, potentially reproductive, behavior.

Identity is a basic life problem having to do with membership in social groups. Isolated individuals are not able to function effectively, or to perpetuate their genes. Identity involves group defense, cooperative hunting, social communication, social signaling, and collective consensus. All species of animals must recognize individuals of their own kind and specific other individuals. Identity is a problem of two opposed primary emotions, *acceptance* (taking in, incorporating) and *rejection-disgust* (expelling). Thus on the social level, the problem of acceptance/rejection concerns who is to be accepted as a member of the species and the group. Of course, identity's emotions, acceptance and disgust, also have to do with one's body (e.g., with food) as well as one's social group membership.

Hierarchy is the "vertical" dimension of social life, involving domination, power, authority, leadership, status, and prestige (Schwartz 1981). With social dominance comes first access to food, sex, shelter, comfort, safety, and enjoyable experiences. Some animals are stronger and more skilled than others, a hard fact that all living creatures must face in their everyday lives: the choice is to fight and struggle for dominance and a high status or accept a lower status. The first solution requires positive behavior directed toward overcoming and resistance to obstacles, which defines *anger*, whereas the second involves *fear* and withdrawal. "Anger" and "fear" refer to subjective feelings but more generally to motivational states underlying the behaviors of fight and flight.

Territoriality also is a basic problem of life. At least for animals as complex as reptiles, birds, and mammals, each animal must separately or collectively establish a territory that "belongs" to it or the group on some level, a safe place that provides nourishment and shelter. A territory is in many instances defined by its boundaries, which every creature marks in some way. Command and control of one's territory require *exploration* and an ability to plan, monitor, and anticipate. Opposed to the opening of territory through exploration is "orientation," with its implied *surprise* and loss of control, as one's boundary is penetrated and territory violated.

From Plutchik and MacLean to Scheler and Fiske: The Four Elementary Social Relations

There is a limit to the usefulness of the models of Plutchik, MacLean, and Chance. As a classification, Plutchik's model is useful for distinguishing primary emotions and constructing definitions of secondary emotions, even though several of his definitions are problematic (see TenHouten 1996, 1999a, 1999b). The major, galling limitation of these models, however, is their *sociological emptiness*. We live in a rich and complex social world, and what we do in this world is much more than one might infer from a reading of Plutchik's and MacLean's work. This suggests that the four-dimensional model needs to be expanded in such a way that it is given *social content*. A corrective can be found in the work of Scheler (1926) and, more recently, Fiske (1991).

Scheler (1926) conceptualized four elementary forms of sociality, paired under two larger principles: (1) kinds of being with one another; and (2), the kind and rank of values in whose direction the member persons see with one another. Scheler himself conceptually unpacked the first, informal level of community organization, seeing that it contained two elements, *identity* and *life community*. He saw formal society as being based on the other two kinds of social relations, *rank* and *value*, corresponding to the institutional domains of politics and economics. There is a full conceptual continuity between Plutchik, Chance, and MacLean—with their identity, reproduction/temporality, hierarchy, and territoriality, on the one hand, and Scheler—with his identity, life community, rank, and value, on the other hand.

Fiske (1991) has identified what he claims are *the* four elementary forms of social life, described as equality matching (EM), communal sharing (CS), authority ranking (AR), and market pricing (MP). Fiske—with no citation to Plutchik, MacLean, or Scheler—has essentially replicated Scheler's notion that there are four basic things we do in the social world.

Fiske's model is simpler than Scheler's but has its limitations. He conflates the meanings of equality matching and communal sharing, for example, as he attributes arithmetical operations to these same two social relations in a way that seems counter intuitive, as he defines communal sharing as "a relation of equivalence" and "common identity," and on this basis links CS relations to the operations "=" and "≠." However, it would be more reasonable, at least on intuitive grounds, to define *equality* matching, and not communal sharing, as a relation of equivalence and identity. The position taken here is that the positive and negative experiences of CS social relations involve the operations "+" and "−," respectively, and that the positive and negative experiences of EM are associated with "=" and "≠," respectively. Fiske largely ignores the *valence* of these social relations, according to which they all can either be negative, positive, or neutral. Moreover, he offers no model of relations between them, nor does he group them or model them as pairs of opposites and complements, as was done by Scheler, and as is done here, but argues for their importance and is content to list them as a set of elementary social relationships.

Scheler argued that the social relations have historically changed with respect to which one is hegemonic in *different stages of human social evolution*: he claimed that the four social relationships have emerged, in the sociohistorical development of civilizations, in the order of identity (in the egalitarianism of primitive, tribal society), life community, rank (with the emergence of political power and the state, the will to power of feudal society), and last, value (a phase of modern capitalism, in which economics becomes the driving force of society). Fiske (1991: 224), however, focuses on the order of developing involvement in his four social relations *in the individual person*, and he proposes a different order, as he sees CS first, followed by AR, EM, and then MP. Itani, were he to use Fiske's terminology, would see EM as the direct result of the negation of AR, which is consistent with Fiske but not with Scheler. These two orderings could both be correct, for they address different levels of analysis; no position on this issue need be taken here.

Fiske argues that the full range of human activity—from participating in religious rituals, to arranging a marriage, to deciding how to fight a fire—is structured in accordance with four fundamental models, four kinds of social relationships. Much of Western social and economic thought has focused on rational choice, agency, and socioeconomic self-interest. Fiske concedes that this principle of "market pricing" is indeed important but correctly insists that noneconomic models also are available to social actors. Because people are naturally sociable, in addition to competing with each other, as in market-pricing social relations, people often choose to share informally and cooperatively (communal sharing) on the bases of family and friendship, to exercise and/or defer to authority (authority ranking), and to balance and distribute things and resources equally (equality matching).

Plutchik's model (table 8.1) is helpful with respect to this issue, because the behaviors he associates with the positive and negative experiences of temporality are "gain" and "loss": consistent with this, the newborn baby is apt to be called "the little *addition*" to the family, and when a family member dies, the family size has been reduced by one, from n to $n - 1$, so the family is in need of reintegration *minus* the one who has died. This analysis suggests that there exists some level of conflation of the concepts of CS and EM, as described by Fiske. The reader is simply referred to Fiske's (1991) book and subsequent journal articles (e.g., Fiske 1992) for his rationale, and to chapter 14 for the way in which the concepts are operationally defined in this work.

From Temporality to Communal Sharing

Recall that for Plutchik the problem of temporality has associated with its positive pole, joy, and with its negative pole, sadness. The positive pole of Plutchik's temporality, reproduction, contains the key idea of communal sharing. Fiske explains that CS has to do with the family, kinship, and human reproduction, arguing that features of CS, captured by English words with a common root ("kind," "kindness"), "… come from an Old English word meaning birth, nature, race, family. All three words are derived from the same Indo-European root, 'ginΩ'—meaning to give birth or beget—from which are derived a family of closely linked words: genus, gender, generous, generate, genitor, gentle, genuine, congenital, nation, nature, and innate (Morris 1970)" (Fiske 1991: 13). Communal sharing, Fiske adds, "… is a relationship based on duties and sentiments generating kindness and generosity among people conceived to be of the same kind, especially kin" (Fiske 1991: 13). Thus the basis of communal sharing is human reproduction, giving birth, and begetting. The human social institution specifically designed for sexual reproduction is the family, so kinship is also implied by the notion of reproduction. In communal-sharing social relationships, people have a sense of solidarity, unity, belonging, and social cohesion: they think of themselves as being all the same in some significant respect, not as individuals but as "we." The scope of CS is wide and can be participated in by unrelated people who are oriented to membership in a particular group.

In CS, reciprocal exchange means that people freely give what they can and take what they need. This was a key notion in primitive communism (or, if you prefer, communalism): "From each according to his abilities; to each according to his needs." To attain distributive justice, corporate use of resources is regarded as a commons, without concern for how much any single person uses. Each contributes what he or she has, without keeping track: "What's mine is yours." Work is treated as a collective responsibility, and people are not divided by specialized tasks. In CS, natal lands received from the ancestors are held in trust for posterity. Land is used corporately—as a commons. Decision making is based on consensus, unity, and a sense of the group. There is a positive value placed on conformity and a desire for similarity in thought, action, and posture. Learning takes place by observation and imitation. Under CS, the group shares a sense of unity, of shared substance (e.g., "blood," "kinship").

From Identity to Equality Matching

"Identity" in Plutchik's sense can be generalized into what Fiske calls the social relationship of "equality matching." The EM principle exists in a wide range of societies. Malinowski (1926: 26) found Trobriand islanders' social behavior "based on a well-assessed give-and-take, always mentally 'ticked off' and in the long run balanced." Thus, for instance, a coastal village delivered fish to an inland village, receiving vegetables in return, after the harvest. Such exchanges occur even between villages at war with each other. EM can exist on the level of turn taking, in which each person in a group performs the same act in a temporal sequence consistent with latent social norms. Equality matching exists as in-kind reciprocity, in which each person gives and gets back what she or he views as substantially the "same" thing; in exchange, there is a balanced and in-kind reciprocity, with appropriate temporal delay. As distributive justice, EM means an even distribution of valuable objects and things so each person receives roughly an equal share; to each the same, regardless of needs or usefulness.

From Hierarchy to Authority Ranking

Plutchik recognizes that hierarchy is a fundamental problem of social life. There is virtually no conceptual distance between Plutchik's "hierarchy" and Fiske's "authority ranking." Authority ranking is a transitive, asymmetrical relationship of inequality. Those with a high rank are regarded as more important than those with a low rank. High-ranking people control resources—people, things, capital, land, and so forth. A basic form of AR was slavery and bondage, where the superior appropriates the will of the subordinate, who also does not control the objective conditions for his or her own labor. A political system based on AR is apt to be an authoritarian dictatorship or to have some other totalitarian form. Distributive justice means that the higher a person's rank, the more choices and resources that he or she receives. Groups have a clear leadership structure and are hierarchically organized. Social identity is defined in terms of superior rank and prerogative, or inferiority and servitude. The positive and negative experiences of AR are associated here, and in Fiske, with the operations ">" and "<."

From Territoriality to Market Pricing

Territoriality is an organizing concept in ethology describing natural behavior oriented to the control of, possession of, use of, and defense of a claimed space that is deemed necessary for survival. Territory is "free" insofar as it affords opportunity for idiosyncratic behavior directed to boundary creation (Plutchik's "exploration") and boundary defense (Plutchik's "surprise").

The analysis of territoriality suggests that the complex and multilevel spaces and places we occupy are closely linked to social relations having to do with resources and valued objects and situations. Territories are valuable resources, necessary for life itself and for its secure enjoyment. The notion of human territoriality must, for purposes at hand, be further broadened to include all forms of possessions, physical and symbolic capital; crystallized energy in the form of money; commodities that can be displayed as a demonstration of wealth; and status, rank, prestige, and privilege. Much of human history is in large measure an account of efforts to wrest territory from others and to defend territory from the claims of outsiders (Hall [1959] 1973: 45).

Market pricing is a relationship mediated by values determined by a market system. In these relationships, people denominate value by a single universal metric, typically price, but also by linear and clock- and calendar-based time. People value actions, service, and products according to the rates per unit of time at which they can be exchanged for other commodities. Market pricing has to do with how people buy and sell commodities. Participation in a market requires quantitative reasoning—addition and subtraction, zero, an additive identity, subtraction, and one distributive law that requires the algebraic operations "$*$" and "$\div$" and attains a ratio level of measurement. Prices, wages, rents, interest, and so forth all have a comparable metric value and therefore exist at the ratio level of measurement, so that any commodity can be bought, sold, or exchanged.

THE SOCIAL *QUATERNIO*

A *quaternio*, as defined by von Franz (1974: 127), is a dynamically related double polarity. This definition is expanded here, so a *quaternio* requires four elements that can be conceptualized both as two pairs of *opposites* and as two pairs of *complements*. This is exactly what we have in the logic of the social relationship model developed in this chapter. The pairs of opposites are CS $\odot$ MP and EM $\odot$ AR, where "$\odot$" means "is the opposite of." The pairs of complements are the logical intersections EM $\cap$ CS, defining hedonic society, and AR $\cap$ MP, defining agonic society, where "$\cap$" means "and." It should be emphasized that this model of relationships between the elementary forms of sociality is not presented by, nor is it endorsed by or agreed to by, Fiske, but is rather an original formulation for which the author takes full responsibility.

Communal sharing and market pricing are *opposites* in the following sense: in communal-sharing relationships, the worldview of the collectivity is internalized in the mind of the individual, but in market-pricing relationships, the individual acts in his or her own self-interest in society yet maintains separation from society. Thus CS and MP represent *opposite mappings of the individual and society*. They also are opposite in emphasizing *communion* and *agency*, respectively. Bakan (1966) describes agency as "… the existence of an organism as an individual, and communion for the participation of the individual in some larger organism" (15). Agency is manifested in self-protection and self-expansion, communion in being at one with others. Agency is competitive, and communion is cooperative. Agency stresses aloneness, isolation, and separation, and communion stresses openness, interdependence, and union.

Equality matching and authority ranking also are opposites. This is true even on the level of logic. Recall that identity, Fiske notwithstanding, involves the algebraic operations "$=$" and "$\neq$," and that hierarchy involves "$>$" and "$<$." According to Itani, the *conditional* equality of hedonic society comes about as the negation or setting aside of hierarchy; accordingly, if it is not the case that $A > B$ or $A < B$, then $A = B$. Equality matching and authority ranking represent opposite human tendencies, to make things unequal and hierarchical, or to set hierarchy aside and attain a conditional equality between people (and other higher primates).

It is proposed that there also are two complementarities. We have seen that equality matching, or conditional equality, together with communal sharing, is the basis of an informal community, which defines hedonic society. We also saw that hierarchy and territoriality, which generalize into authority ranking and market pricing, are the basis of a formal, agonic society. It can be claimed here that hedonic and agonic society are not merely qualitatively different but *real opposites*, because their elements are opposites: more specifically, the elements of hedonic society, EM and CS, and the elements of agonic society, AR and MP, are opposites because EM and AR are opposites, and CS and MP are opposites. Thus the logical requirements for a social *quaternio* are satisfied. These relationships are displayed visually in figure 8.1.

In panel A, the CS line segment has the positive experience of CS (CS^+) at one end and the negative experience of CS (CS^-) at the other. The MP^+/MP^- line segment is shown perpendicular to CS^+/CS^- because MP and CS are opposites. Panel B shows the same relationship for the EM^+/EM^- and the AR^+/AR^- line segments. In panel C, the first two panels are superimposed. The resulting figure, a mandala (representing four dimensions in two dimensions), represents the *social quaternio*

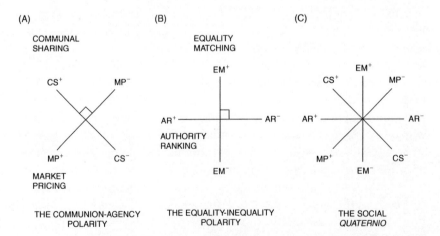

FIGURE 8.1 (A) The communion-agency polarity; (B) The equality-inequality polarity; (C) the dynamic double polarity of opposite elementary forms of sociality, the social *quaternio*.
Source: TenHouten 1999a, figure 3, p. 59.

(TenHouten 1999b: 263–66). The two oppositions are a matter of theoretical reasoning. The validity of the two complementarities that result in a hedonic and an agonic society, in contrast, is an empirical question, which can now be dealt with by examples from Aboriginal and Euro-Australian social behavior and also with empirical data, discussed in chapter 15.

Summary and Discussion

This chapter has shown a remarkable continuity regarding the work of Chance, Plutchik, MacLean, Scheler, and Fiske. The mere fact that a small group of scholars arrived at the same general conclusion is, of course, no guarantee of its validity, and the social-scientific literature is replete with instances of far larger sets of scholars who have agreed upon models of the social world that are, in retrospect, mistaken or too vague to be investigated empirically. Consider, as a prime example, the Parsonian model of "pattern variables" that enjoyed a few decades in the sun as the dominant theoretical paradigm in North American sociology, before being relegated to the dustheap of science past. When confronted with the present conceptual model of four social relations packed into a social duality, many interested readers will immediately object, complaining that things cannot be that simple, and rightly pointing out that there is nothing magical about the number four. However, the enterprise of science is nothing more than a search for simplicity of explanation, and we all know that Mother Nature has nothing to hide. If this scheme is indeed too simple, it can be modified in (at least) three ways: (1) by eliminating one or more of the four criteria, but this seems unreasonable and would, for example, eliminate the primary and secondary emotions that are known to exist; (2) by adding one, two, or more essential problems of life to the model, which is a definite possibility that would not be logically inconsistent with the present formulation, but the reader is urged not to hold his or her breath until such conceptual elaboration comes to pass; and (3), it has been argued by postmodernists, social constructionists, and other that there are no *universal* problems of life, therefore no primary emotions (e.g., as argued by Ortony and T. J. Turner 1990), and that there are no *elementary* forms of sociality because human culture and society are largely unconstrained by biology and have been socially constructed in a somewhat arbitrary manner. This argument could then be extended to the claim that there are no general modes of information processing structured by brain design (e.g., as argued by

Efron 1990 in his refutation of dual-brain theory), and therefore that there are no elementary forms of time consciousness. There is absolutely no hope of predicting kinds of time consciousness that do not exist from kinds of social relationships that also do not exist, so the ultimate test of the present theory is an empirical problem, and the kind of validity needed is *predictive* validity. Thus the results of a study designed to test the theory, presented in chapter 15, will answer these criticisms.

Chapter 9

Natural and Rational Experiences of Time

Four kinds of social relationships have now been explicated, as have four kinds of time consciousness. In this chapter, we will briefly examine the logical structure of these small social and temporal models and then construct an isomorphic model for the four most general kinds of information processing made possible by the human brain.

THE TEMPORAL *QUATERNIO*

Just as our social relationship model consists of four elements that comprise two pairs of opposites and two pairs of complements, the present model of time consciousness has the same logical structure and can on this basis be referred to as the "temporal *quaternio*." It has been shown that patterned-cyclical and ordinary-linear kinds of time are opposites, because the seven aspects of patterned-cyclical time, P1–P7, when turned into their opposites, L1–L7, result in a full description of ordinary-linear time. When we examined the more general modes of thought on which patterned-cyclical and ordinary-linear time are based, the gestalt-synthetic and logical-analytic, we further discovered that they, to some extent, were based on opposite principles of *simultaneous* and *sequential* information processing, respectively. It also has been proposed that immediate-participatory and episodic-futural kinds of time consciousness are opposites, on the ground that the former requires a temporal *compression* of the present and future into the present, whereas the latter requires a temporal *stretching* of the present into both the past and the future. When we examined the biological bases of these two kinds of time, the participatory and the episodic information of the posterior cortex and the frontal lobes, a "dialectical" relationship was found between them, which Laughlin (1988) calls the "prefrontosensorial polarity principle."

Just as there is opposition, so there is also complementarity. There is a broad complementarity of immediate-participatory and patterned-cyclical time among the Aborigines. Consider an example from Sansom (1980), who describes the process of reaching consensus among the fringe dwellers of Darwin. Participation in the construction of a consensus, the "word," takes the form of progressive recruitment of people to reveal the details of the story, thereby bearing witness and engaging in the embodied work of righting and straightening a "blaming" story. It would not be unusual for a primary witness to speak for hours when contributing to the story. Such a story is considered made "when all the righted details have been fitted together" (Sansom 1980: 130). The blamed person will submit to the story and endure the social isolation of a person intentionally made lonely. The *gestalt* consisting of the entire social group's determinations needs to be *au courant*, up to the minute, so people coming and going need to be updated about important events. The objectified definition derives from the way in which "the determination in a warrant establishes a relationship between the present and the near but determined past" (Sansom 1980: 133). Thus there develops a shared consensus, a gestalt, that is built up and that must be up-to-date, in the present. This temporal experience is constitutive of *natural time*.

The dominant value system of modern, advanced societies includes universal education, a belief in equal opportunities for economic success, an emphasis on achievement, and a striving to "get ahead." Struggle is expected for influence, power, prestige, and wealth. Contemporary education facilitates the attainment of a high status in modern, capitalistic societies. Much of the world of commerce, such as banks and insurance companies, is based on a striving for the future (Duncan 1980: 53). What is required of the successful individual's personality is a sense of ambition, initiative, and competitiveness. Very central to the reproduction of this value system is to learn early the importance of time in the struggle to get ahead. Writing about the American version of this system of values, Halpern (1973) says this of the ambitious boy:

> So in his early years he was taught that it was important not to waste time. He also learned the need for punctuality, and many other concepts associated with the matter of time. Similarly, he was impressed with the need

to plan and organize his activities, and in this connection, time as well as many other attitudes and attributes were brought into the picture. (29)

Thus striving for future economic success is linked to the mastery of ordinary-linear time and, more generally, the attainment of logical-analytic reasoning skills, the result being an immersion in agonic society and the accomplishment of a rational time orientation. There is a complementarity of linear, clock time, and episodic-futural time. Heidegger ([1927] 1962) realized and made clear, as we have seen, that future-directed action taken over time is apt to be described by members of modern societies in terms of *ordinary* time, so the accomplishment of objectives comes to be seen as being realized at some point in ordinary time in the future, perhaps as some date on the calendar marking a deadline. Heidegger, to his credit, has not gone entirely beyond and fully negated the concept of ordinary-linear time insofar as at least an approximate beginning and an approximated future end typically bind the temporal stretch, possessing the characteristic of "dateability." Thus ecstatic-futural, primordial temporality is, in the everyday, modern world, described, either directly or indirectly, in a context of linear time. The tasks of life are never finished, and our goals, as we meet them, are talked about and scheduled in terms of clock- and calendrical-based time. There is thus, at least potentially, a veritable fusion of ordinary-linear time and episodic-futural time, which together can potentially result in a higher-level, *rational* time awareness. There also is thus a fundamental complementarity of episodic-futural and ordinary-linear time, crystallized in the attainment of rational time awareness. So to plan for the future is to work out priorities for actions and initiatives, which requires the notion of linear time (Ferrarotti 1990: 74). Future orientation thus employs linear time as an instrument of measure and control.

The Biological Basis of Natural and Rational Time

Opposition: Left and Right Hemispheres

Right hemisphere (RH) dependent processing and Left hemisphere (LH) dependent processing have been widely seen as opposites (Bogen 1977). Each is specialized for *synthesis* and *analysis*, opposite cognitive processes. Each is specialized for *simultaneous* and *sequential* processing, also opposites. Both hemispheres control and receive information from the opposite side of the body. In the case of vision, the left half visual fields of both eyes inform the RH, and vice versa. Levy (cited in F. Turner 1985: 70) has argued that the linguistic capacities of the LH provide *a temporal order for spatial information*, which is the opposite of the melodic capacities of the right brain, which provides a *spatial order for temporal information*. And in an important statement, Bogen (1973) has proposed that the RH maps the world as a subset of the self, and that the LH maps the self as a subset of the world. There also is evidence that when information is transferred from one side of the brain to the other, it is somehow inverted, so that one hemisphere's representation is a "mirror image" of the other's. For example, a right-handed person trying to write with the left hand is apt to discover an advantage in mirror writing. The hypothesis of that visual information about the world represented in mirror-image form was first developed by Orton (1937), who had observed that dyslexic children often confused left from right, had incomplete cerebral dominance, and were prone to invert letters and words and to write in mirror-image form. Orton's explanation of dyslexia is no longer accepted, and the phenomena of dyslexia are far from resolved. It is now known that children with such learning difficulties often have a slow development of the corpus callosum, which limits the effectiveness of information transfer from one hemisphere to the other, which tends to disappear as the corpus callosum matures and develops. It also has been found that when a split-brained monkey learns that a particular shape is a correct answer in a binary visual discrimination task, then transfers the learned shape to the other hemisphere via the corpus callosum, the choice of the other hemisphere is apt to be the mirror image of that shape, if such a shaped object is presented as an alternative. In this experiment, Noble (1968) used two stimuli that were left-right mirror images of each other (figure 9.1, panel a). He severed the optic chiasm and later covered one of the monkeys' eyes and taught them to select one of two stimuli. After training, the monkeys were presented the same stimuli with the sight of the trained eye blocked (figure 9.1, panel b), and they eventually showed a preference for the cue that had not been rewarded, the mirror image of the "correct" answer. Noble concluded that the information was reversed during its transfer across the corpus callosum. To test this conclusion, he repeated the one-eyed training (figure 9.1, panel c), except after learning had taken place, he split the corpus callosum before presenting the pair of stimuli to the untrained eye (figure 9.1, panel d). Here the trained hemisphere was rendered unable to transfer the mirror image of what it had learned, and the monkeys were able to select, by other ways, the cues they had originally learned.

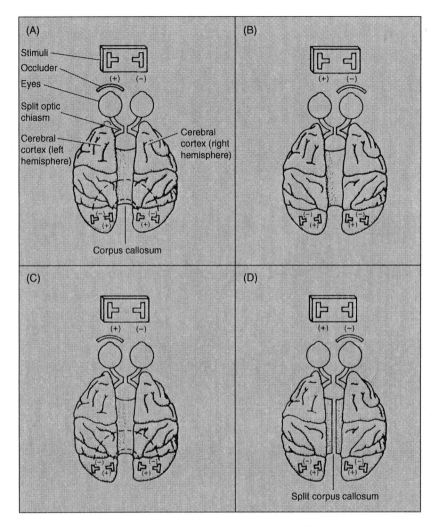

FIGURE 9.1 Information transfer in mirror-image form in Noble's 1968 experiment with monkey subjects.
Source: Corballis and Beale 1971, p. 98.

OPPOSITION: POSTERIOR AND ANTERIOR CORTEX

Pribram (1981) names the overall mode of information processing of the posterior cortex and frontal lobes "participatory" and "episodic," respectively. It has been argued that the immediate-participatory and episodic-futural kinds of temporal experience are key aspects of these general modes of information processing. Immediate-participatory time consciousness requires a concentrated effort to mindfully focus on what is happening in the present, and in this sense can be said to be "on-line." Episodic-futurality, in contrast, requires a mobilization of resources from the past, most notably a model of the world, so that effective acts and actions can be carried out in the future, which takes the thinker "off-line." Participatory time processing and episodic time processing also are seen as opposites insofar as the former involves temporal compression, and the latter involves temporal stretching.

COMPLEMENTARITY: RIGHT HEMISPHERE AND POSTERIOR CORTEX

There is a strong complementarity between the participatory processing of the posterior cortex and the gestalt-synthetic information processing of the right hemisphere. *Both are immersed in the present.* The tertiary-level closure and pattern-recognition processes of the RH are simultaneous, just as is sense perception. In the development of the human, the right hemisphere develops earlier than does the left hemisphere, which is not a new phenomenon with humans but rather has a long evolutionary history and exists in other species as well (Geschwind and Galaburda 1987: 44–45). It is of advantage to the individual for the RH to develop first because of its specializations: it is dominant for a wide variety of spatial functions, including the orientation of the body in external space. The RH plays a major role in the cognitive representation of emotions, both in the subjective experience of emotion, both of self and of others, and in the expression of emotion. The RH is, Geschwind and Galaburda add, of predominant importance in attention, which is closely linked to the emotions and to concentration on external stimuli. Thus the RH, from the very beginning of life, plays a special role in mental activities essential for personal survival, so there is a close connection between perception of the external world and the RH's vigilant synthesizing of this information. In the early stages of life, survival itself depends on the RH's ability to concentrate on the basic problems of life—to identify the mother's smiling face; to communicate needs, desires, moods, and feelings nonverbally; to demand and receive attention, care, and love. At this stage of life, the RH is vitally important, acting not so much as an organ of thought as an organ of survival. As the child ages and develops language and analysis, the LH will become increasingly important and, especially in a modern society with a formal educational system and an economy that rewards clarity of thought, will, as a matter of course, become dominant. But the unity of the posterior cortex and the right hemisphere will continue to enable us to act naturally and, importantly, to develop a natural time consciousness. The left hemisphere, in contrast, is destined to look not backward but forward, to the frontal lobes and the development of *linguistic analysis*, of reason and rationality in general, and of a rational experience of time as an essential component of a future-oriented Reason. The right hemisphere, in the right-handed adult, ordinarily has only a limited involvement in language. Ordinarily lacking an active language system, it functions mainly at the perceptual-motor level, a simple fact that suggests a close linkage of the participatory processing of the posterior cortex and the integration of such perceptions by the right hemisphere.

Complementarity: Left Hemisphere and Frontal Lobes

There exists a complementarity of frontal-lobe dependent, episodic functioning and the logical-analytic, language-enabled functioning of the left hemisphere. The frontal cortex, as we saw in chapter 7, gives the human species great conceptual powers that depend on language. The frontal lobes are responsible for initiating a motivation to produce speech (and writing), guiding and controlling search activities associated with purposive reasoning, and formulating prelinguistic "semantic graphs" of ideas. Patients with severe frontal-lobe damage might be able to talk in grammatically correct sentences but lack the motivation to do so. They have difficulty sustaining interest in problems and projects so complex that their solution requires the use of language. Luria associated the posterior area of the frontal cortex, especially the *pars opercularis*, with the function of translating semantic graphs of ideas into inner speech. While ideas are organized non-sequentially, the structure of inner speech is syntactically and sequentially organized. Luria (1982), in this connection, posits a dialogue between ideas (semantic graphs) and inner speech, one that requires translation between two different neuronal "languages." Drawing on the work of Vygotsky, he claims: "Thought is *completed* in the word ... [and] thought itself is formed with the help of the word or speech" (152). Thus the language of inner speech structures the content of thought, and in the same process, it modifies thought, So there is interdependence between the episodic information processing of the frontal lobes and the logical-analytic, linguistic processing of the left hemisphere. Perhaps the most obvious structure involved in this connection is the superior longitudinal fasciculus (or arcuate fasciculus), a veritable "cable" that connects the posterior area for the comprehension of speech, Wernicke's area, in the left temporal lobe, and Broca's area, in the left frontal lobe, for the production of speech. Based on this evidence, it is proposed that episodic-futural and ordinary-linear time consciousness finds a potential unity in *rational* time consciousness, which has as its infrastructure this frontal-lobe/left-hemisphere connectivity and the integrated mental functioning just described.

The pre-motor and motor areas of the left cortex are important not only for speech production but also for reasoning. These cortical areas are located along the top of the left frontal lobe, merging posteriorly into the motor or pre-central cortex, which is considered anatomically intermediate between the frontal and posterior cortices. These motor areas enable us to articulate individual sounds and also provide the "kinetic melodies" (Luria 1982: 15) or "rhythmic structure" (Brown 1982) necessary for linking words in the proper, linear order. Thus while ideas are nonlinear, the translation functions of the posterior portion of the frontal cortex are able to linearize ideas, to represent them with sequentially organized utterances (Boyle 1985: 68). Patients with damage to this area of the frontal cortex have difficulty organizing their ideas in a meaningful temporal order. Without this melodic/rhythmic structure, words are apt to be deleted or used in the wrong order, confusing both the listener and the patient. Thus it is not the case that speech is formed separately and then transmitted to the motor areas for expression; instead, formulation of the motor flow of the intended utterance is part and parcel of the circuit necessary for an idea to be verbally articulated (Boyle 1985: 69).

Mental dialogue, then, is shown to be a process in which both frontal and posterior areas of the brain participate. The translation of ideas into inner speech leads to a rhythmically articulated, grammatically elaborated structure, which can be represented to the frontal cortex as higher-order perceptions. Languages contribute to this mental dialogue by making available their vocabularies for encoding meanings and their own logical structures for elaborating and sequentially ordering meanings. So mental dialogue involves "talk" between the front and back of the left hemisphere. Because of the great importance of language for intentionality, intentional mental processes cannot be strictly "localized" in the prefrontal cortical areas of the brain. Research by Brown (1982) suggests that cortical areas *posterior* to the prefrontal cortex can dominate mental dialogue. He found that the greater the level of electroencephalographic activity in the motor cortex, the greater the subjects' feeling of intentional control over their

thought processes. Thus when we have in our mental c
Heidegger called an "anticipatory resoluteness" to realize som
of affairs, there must, it would seem, be a *motor* plan of actio

CULTURAL UNIVERSALS: A LOOK AHEAD

The four elementary forms of sociality have a biological basis and an evolutionary history and, on this basis, can be seen as cultural universals. But the same cannot be said for the hedonic and agonic forms of community and society. That they are not universal follows from the simple fact that the hedonic community characterizes apes and the agonic society characterizes monkeys, but neither class of primates possesses both forms of social organization. A similar argument holds true for time consciousness. While the four kinds of temporal experience have been proposed to be culturally universal, on the grounds that the four modes of information processing, of which they are aspects, exist in all normal brains, the same is not true of natural and rational time experience, which might or might not emerge as cognitive structures in the mind of any individual human being. With this in mind, we now turn to the first four propositions of the theory, according to which involvement in each of the four elementary forms of sociality predisposes the member of society to a cognitive structure that emphasizes one of the four elementary forms of temporal experience. Chapter 10 considers communal sharing and patterned-cyclical time; chapter 11, equality matching and immediate-participatory time; chapter 12, authority ranking and episodic-futural time, and chapter 13, market pricing and ordinary-linear time.

Chapter 10

Communal Sharing and Patterned-Cyclical Time Consciousness

Communal sharing, as described in chapter 8, is a sociological generalization of what Plutchik ([1962] 1991) calls "temporality." Temporality is the limitation and erosion of time, and the family is its universal social institution. The basic process of temporality is birth and death, with its positive pole of sexual reproduction and its negative pole of reintegration of the family and community following the loss or death of family members. Communal sharing also applies to larger collectivities, such as communities, clans, and tribes, as it is a relationship based on duties and sentiments generating kindness and generosity among group members. Communal-sharing relations are validated through connection to the past: this is the core meaning of tradition and the basis of ritual. The proposition of this chapter is that the higher the level of *positive* experience of communal-sharing relations for societal members, the more involvement there should be in a patterned-cyclical time consciousness.

The focus of this chapter will be the ethnographic analyses of our case-study culture, the Aborigines, which will link communal-sharing social relationships to patterned-cyclical time consciousness in general and to each of the seven aspects of patterned-cyclical time consciousness in particular (P1-P7, table 5.1). Communal-sharing relations, of course, exist in every human society, but cross-cultural differences exist. Communal living, it is proposed, is *central* to the social organization of hunting-and-gathering and myriad other kinds of traditional, oral, indigenous, and pretechnological societies, and the Australian Aborigines are no exception to this generalization. The traditional Aborigines were, and to some extent still are, in everyday life, "broken up into small groups of mixed local-descent membership; and they relied on cooperation through the spectrum of social relations framed in kin terms" (Berndt and Berndt 1980: 9). Even among detribalized, urban, and suburban Aborigines, the extended family retains centrality. It could well be the case that in suburban and urban areas, where access to language area, tribe, clan, and community of origin, and the way of life provided, is no longer available, the family and the extended family have an even higher level of centrality and importance.

The proposition is not that Aborigines are higher on communal-sharing relations than are their Euro-Australian controls. In fact, the pathology and breakup of families resulting from the government policy of taking children away from Aboriginal parents, along with the tangle of pathology and its associated psychological sense of brokenness resulting from grinding poverty and dispossession, have led to, in family and community life, a decrease in the positive experience of communal sharing and an increase in the negative experience of communal-sharing social relations. If this is true, the result should be a reduction in a patterned-cyclical time orientation, because, by hypothesis, it is only the positive experience of communal sharing that is predicted to contribute to patterned cyclicity. At the same time, we can expect that this loss of a traditional time orientation would be mitigated by the Aboriginal tendency, reinforced by their culture, to retain a spatial as opposed to a temporal orientation, discussed in chapter 6.

The Pintupi term *walytja*, as interpreted by Myers (1986: 48), captures much of the meaning of communal sharing. Aboriginal societies have highly elaborated classificatory models of kinship that incorporate all individuals into a system of kinship. The classificatory elaborations are known to reflect Aboriginal concerns with constituting a sociality between the local group and a larger tribal membership (Maddock 1972; D. H. Turner 1980). Cultural emphasis is given to the temporal notions of "continuity" and "permanence" (Stanner 1968). Every named place is part of an encompassing structure of places, and rights to the sacred places do not function simply to exclude people without these rights, but, rather, place individuals in a larger order of social relationships. "Every ritual related to land," Myers incisively observes, "is an instantiation of a *societal whole*.... The organization of ceremony, requiring participation of others from far away, provides one way of constituting Pintupi society as a whole" (1986: 180, emphasis added). The system of "classificatory kinship" is based on the participation of others beyond the immediate family in the production of individuals as full members of society. Here "kinship" is a system concerned with the

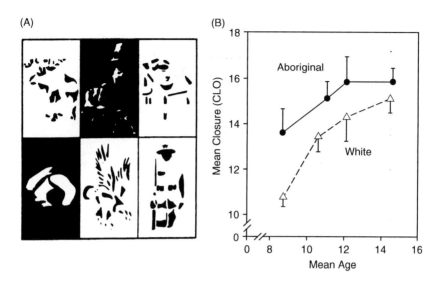

FIGURE 10.1 (A) Selected items from the Closure 79 test of visual closure; (B) Mean closure performance for Aboriginal and Euro-Australian children after partition of both groups into four age levels. Error bars are +/− standard error of the mean.
Source: TenHouten 1985, figure 3, panel A. Reprinted with permission of Gordon and Breach, science publishers.

(living things), there was no difference at the youngest age level, but for the older age groupings the Aboriginal children appeared to lag behind their Euro-Australian controls (figure 10.2, panel A). It also was found that in response to both the seventeen Similarities (SIM) questions, and an original set of seventeen other questions (NEW), Aboriginal children were significantly more apt to manifest leftward conjugate lateral eye movements, or lateral saccades, indicating relative right-hemispheric activation (TenHouten 1986). These performance differences and cerebral activation differences both point to an Aboriginal right-hemisphericity in comparison to their Euro-Australian controls.

Fusion of Past and Present and Communal Sharing

When communal sharing is interpreted on the level of temporality, it is the past that is reproduced through ritual and tradition. Communal sharing supports a kind of time consciousness that links the past to the present and

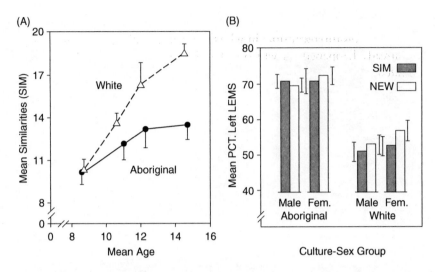

FIGURE 10.2 (A) Mean Similarities performance for Aboriginal and Euro-Australian children after partition of both groups into four age levels. Error bars are +/−1 standard error of the mean; (B) Mean percentages of left lateral eye movements in response to Similarities and NEW items, by culture and sex. Error bars are +/−1 standard error of the mean.
Source: Panel A from TenHouten 1985, figure 4A; panel B from TenHouten 1986, figure 2A. Both reprinted with permission of Gordon and Breach, science publishers.

is stimulated by ritual. For the cosmologies of Australian Aborigines, there is for every tribe an elaborated cosmology, characterized by narratives, songs, dances, and other artful renderings of the exploits of cultural heroes during a primordial, creator-being time that accounts for the development of the earth, for their distant ancestors, and for themselves living in the world. In the sacred narratives—songs and stories—the riddle of time is the riddle of the "beginning." As van der Leeuw explains, "In the beginning lies the whole past. The beginning is the past. Yet we say that we begin something, that we make a new beginning. And we call the long list of such beginnings, time" (1957: 325). To say we are temporal means that we can neither grasp nor hold fast to a point in time or to our own existence. The dividing lines of our lives are singular events. "Time is full," van der Leeuw writes, "and its content is the myths—that is, the forms of life that lie heaped up for all time in the womb of the unconscious" (1957: 329).

The existence of the world is not guaranteed but must rather be begun anew through the enactment of ritual, this ritual reenacting the exploits of creator beings. This idea of a primordial time is not a fixed date in the distant past but is rather a connection of the past and the present, and a manifestation of the sacred, inner reality with the current reality of enacting the ritual and the story it tells. For van der Leeuw, this kind of temporal experience is "a prototypical time in which everything of cultural significance has already happened, in which today does not begin anew but is only repeated; primordial time is creative; it creates what happens today. And this is brought about through the repetition of its myth" (1957: 332). Thus myth creates time, giving it content and form. Time is dualistic, split into two levels of reality. This duality of being, however, does not lead to alienation but rather to unity and—through a protracted effort and successive initiations into the sacred-secret knowledge of the tribe—to a *sense of totality* brought about by the feeling of oneness with the Dreaming and by the psychologically pleasing experience of synthesis.

Lévi-Strauss (1965), in his early structural analysis of the Oedipus myth, emphasizes both its logical and temporal structures. Myth in general, he contended, has three properties: (1) the meaning of myth resides in "bundles" of relationships between symbolic elements; (2) there are two analytically distinct aspects of the presentation of these bundles, the difference of which is their mode of temporal extension: one aspect, analogous to Saussurean *langue*, possesses a time that is reversible and synchronic; the other, analogous to *parole*, has a time that is irreversible and diachronic; and (3) the unique and essential character of myth combines or fuses diachronic and synchronic temporal extension, in a synthesis that Lévi-Strauss calls "mythic time." Myth, Lévi-Strauss explains, purports at once to be "a non-reversible series of events ... belonging to the past ... and an everlasting pattern which can be detected in the present ... social structure and which provides a clue for its interpretation" (1965: 85). Thus a fusion of past and present is interior to the very structure of myth. Moreover, "a myth always refers to events alleged to have taken place in time: before the world was created ... long ago. But what gives the myth an operative value is that the specific pattern described is everlasting; it explains the present and the past as well as the future" (Lévi-Strauss 1965: 85). What Turner (1969) calls "[t]he structural locus of this fusion of the synchronic and the diachronic" (27) is for Lévi-Strauss not to be found "in the style, its original music, or its syntax, but in the story which it tells" (1965: 86).

T. S. Turner (1969) points out serious flaws in Lévi-Strauss's arguments. To see mythic time as fused, he insightfully suggests, is a misguided attempt to reconcile two incompatible modes of temporality. Lévi-Strauss, according to Turner, correctly sees that a synthesis or creative juxtaposition of the synchronic and the diachronic is the central principle of myth but misunderstands the nature of the problem, for Lévi-Strauss virtually displaces the diachrony of the mythic narrative with the interminable synchrony of the myth and immediately loses narrative sequence. He subordinates sequences to schemata, according to which myths need time only in order to deny their place. But as T. S. Turner (1969: 31) points out, Lévi-Strauss had in his initial theoretical statement claimed that "the essence of myth is inextricably bound up with the 'story that it tells,' i.e., the 'diachronic'…order of the narrative" (1969: 31). And, Turner (1969) adds, Lévi-Strauss is correct in his claim that myths "do indeed provide synchronic models of diachronic processes, but they do this through the correspondence between their sequential patterns and aspects of the diachronic processes they 'model'" (32). By reducing narrative to synchronic paradigms of semantic components, the original object of study—the narrative or tale itself as a coherent whole—is annihilated. The "plot" or narrative structure of myth begins with events that violate the existing natural-moral order, giving rise to a situation where the actors and elements are engaged in contradictory, conflictual, confused relationships, then runs through a series of permutations in which relations are restored. There is thus a dialectical alternation between synchronous order and diachronic, disorderly change (Turner 1969: 33). This alternation creates a *coherent whole*, so the worldview leads to a collective consciousness, a shared set of paleosymbolic images that are at the heart of communal sharing, as in celebrating rites and rituals and carrying out traditions, and the mythology of the heroic past is linked to the present through narrative and other performances.

Discontinuity, Irregularity, and Communal Sharing

In chapter 4 it was argued that, for Aborigines, the punctuation of everyday, mundane time by the time out of time of the formal ritual creates a sense of time that is heterogeneous, irregular, and discontinuous. Because all formal rituals can be classified unambiguously under the heading "communal sharing," it follows that time consciousness in the midst of communal sharing can be further characterized by the aspect of patterned-cyclical time, "irregular, discontinuous, and heterogeneous" (P3). The very discontinuity, irregularity, and heterogeneity of the Aboriginal, totemistic belief system create a situation in which disparate, even contradictory elements, including the primal discontinuity of the sacred and the profane, create the raw materials for a worldview that must be synthesized, much as fragments of a whole are put together in the process of gestalt completion, as argued earlier and illustrated in figure 10.1, panel, A.

Lévi-Strauss (1966: 189) contends that classification, a cognitive capability that allows humans to approach nature in an indirect way, creates a cognitive experience of discontinuity. Durkheim ([1912] 1965) saw the alternation between times of the profane and the sacred as an oscillation of two distinct kinds of time, ordinary life time and sacred time, where emotionally evocative interaction rituals and other social events are carried out with lawlike replication. Each replication provides for its participants the phenomenal experience of inner reality and awareness of the eternal. Through injunction, taboo, interdiction, ascetical practices, and negative rites, this replication both substantiates and separates the sacred and the profane. Durkheim ([1912] 1965) regarded the sacred-profane distinction as absolute, variously referring to these terms as *antagonistic, jealous, hostile*, an *abyss*, and a *logical chasm* (but he also recognized an ambiguity of the two). Irregularity and discontinuity of time can be seen in the rupture of the eternal rhythms of nature—brought about by protracted drought, pestilence, disease, and so on. The reestablishment of the natural-metaphysical order of the world requires, the Aborigines believe, the enactment of rituals based on esoteric information, revealed only to certain mature men, or "men of high degree."

The correlation of community and social relations, however, is problematic for Aborigines, even for those living close to their old tribal ways. Myers (1979) explores this problematic in his study of the Pintupi by a focus on time and emotion. He found that the most salient problem of Pintupi life is the community, which is rooted in the way that the Pintupi articulate the relations among people who live together as family or kin (*walytja*). Aboriginal communities are defined not by physical boundaries but by networks of fragile social relations sustained over time only with intention and effort. Polity does not reflect authority but a jurisdiction of relatedness among intermittently autonomous individuals. Correct behavior in face-to-face interactions among the Pintupi, and among Aborigines in general, is culturally defined. It is important to be able to gauge how, when, and where to express feelings of shame. The Pintupi concern with shame contributes to a presentation of self largely devoid of egotism and selfishness. Instead there is a commitment to relatedness, "which emphasizes the shared goals of egalitarian, closely cooperative kinsmen" (Duncan 1986: 47; also see Myers 1979: 362). Traditional Aboriginal society is in general self-regulating, with shaming used as a control mechanism of thought, word, and deed (Duncan 1986: 52). Equality of effort is supported by the Aboriginal custom of helping followers who are in need and by stressing cooperation.

EVENT ORIENTATION AND COMMUNAL SHARING

Formal interaction rituals, along with a variety of informal happenings, are of the utmost importance for tribal-living Aborigines. Aborigines are not, in general, future-value oriented, but they do look with serious and concerned attention to events that have not yet occurred. Thus we can add to the description of the doing of time in the midst of communal relations the quality "event oriented" (P4). Aborigines are event oriented because of the extraordinary value they place on the sharing of experience (Sansom 1980: 3). The instrument of sharing of experience, and of like-mindedness, is, of course, language. Thus we might expect Aborigines to place great value on the spoken word, and indeed they do. Sansom's fringe-camp Aborigines and Aborigines of the Northern Territory and beyond ordinarily have knowledge of two, three, or more languages. They are, Sansom notes, "highly self-conscious about the use of words; they speak about speaking and have the words to talk about talking and so bring down refined judgments on a speakers' performances. They set great store by what they themselves call 'the word'" (1980: 4). Aborigines, along with members of every other human culture, use words in order to create and arrange their social forms, to bring into being their collective life. The Aborigines' high level of event orientation is an expression of the importance they place on communal life and the striving for like-mindedness. At Wallaby Cross, Sansom (1980: 84) reports, one cannot simply "tell what bin happenin' and give a blow-by-blow account to someone who did not witness the event." There must rather be a public recounting of events before a general audience, with talk at these gatherings governed by two rules: (1) the recounting of details is subject to special and approved reasons, that is, a camp member may not "tell what bin happenin for nothing"; (2) the second rule modifies the first: "If a communal verdict has been formulated to describe the nature or outcome of a completed event, the communal verdict may be recounted without breaking the first rule" (Sansom 1980: 84). The objective is not to "tell details" but to "give the word." Sansom provides the following examples:

(1) "Big Jack bin bash Olive [his wife] for no reason. She sick now";
(2) "Violet bin diblorce Lionel after fightin'";
(3) "Yesterday ol Luke bin miss that promise" [failure to secure a promised bride];
(4) "Prue bin get up before sunrise and cook for that olfella in camp." (1980: 84)

Such verdicts communicate changes and shifts in camp affairs, telling interested people that as a result of a chain of unrecounted events, there is a new state of affairs. Prue, for example, had left camp and her husband months ago and is now performing for all to see and comment upon ("everybody sayin") her resumed status of wife. What is registered is not a résumé of past events but rather the defined happening upon which the subsequent behavior of Prue and her husband is to be based.

Cycles, Patterns, and Communal Sharing

Rose (1992) writes of the Yarralin people that "Temporally, one's life progresses back toward the Dreaming. As generations are washed away, ordinary time is collapsed into the Dreaming" (209). She explains that people's lives expand in terms of spatially located knowledge, but toward the end of life, "people refocus their attention on the country in which they wish to die. The whole of life, one might say, is a great circle from the earth, around the top, and back to the earth" (ibid.). Thus in addition to reincarnation cycles, cyclicity also can be found within the lifetime.

The rhythms of nature—associated with the day, the month, and the year—are for the Aborigine neither separate nor mutually exclusive. As M. D. Young (1988) writes: "Organisms do not respond to the solar and lunar rhythms separately, one after another; we ... reach to them simultaneously, synthesizing them into one response in a continuing present" (30). Insofar as Aborigines tend to have a gestalt-synthetic mode of thought (Grey 1975; Lewis 1976; TenHouten 1985, 1986), it is reasonable to presume that the times of the day and year, as described earlier, are perceived as a simultaneous, constantly updated gestalt.

While great effort has been made in Western and other civilizations to find a logical and quantitative understanding of these cycles of nature, an effort greatly complicated by their numerical nondivisibility (e.g., of the year into months),[2] among tribal-living Aborigines no such logical connections are made. Aboriginal time is associated with the circadian, circumlunar, and circa-annual rhythms of nature.

Quality and Community

In their study of Australian Aboriginal religious categorization and classification, Durkheim ([1912] 1965: 488) and his followers (Hubert 1905;

Hubert and Mauss 1909) developed a concept of "qualitative time" (see Hassard 1990: 2–3), emphasizing the rhythmic nature of human society but also acknowledging the necessity of understanding time on a multiplicity of levels synthesized not by logic or verbal clarification but rather by a qualitative assessment of interdependent natural and social phenomena. Durkheim conceptualized time as a symbolic structure, a collective representation of the organization of society through temporal rhythms (see Isambert 1979; Hassard 1990: 2). He saw the "social time" of extraordinary rituals as a phenomenal experience of the eternal and as a product of collective effervescence. In these rituals, the Aborigine of high degree recreates his own celestial history without pinning down the Dreaming to some arbitrary point in time.

Tribal-living Aborigines place great emphasis on social consensus, in which their collective participation virtually requires every member to verbally agree to a decision. Thus a decision is a collective effort in which new ideas and proposals are introduced incrementally without attention focused on any leader (see Liberman 1985). Durkheim wrote that to make a collective representation, such as the notion of time, "a multitude of minds have associated, united, and combined their ideas and sentiments: for them, long generations have accumulated their experience" ([1912] 1965: 29). The Aborigines' *qualitative time*, in this analysis, is "the sum of the temporal procedures which interlock to form the cultural rhythm of a given society" (Hassard 1990: 3). Durkheim wrote: "The rhythm of collective life dominates and embraces the varied rhythms of all the elementary lives from which it results; consequently, the time which is expressed dominates and embraces all particular durations. It is time in general" ([1912] 1965: 490).

Long Duration and Traditional Community

The individual's reflective self must endure by opposing its changeability in time to its permanence. The experience of patterned-cyclical time thus finds its fullest realization in the awareness of duration, an awareness of the coexistence of the past, present, and future and the experience of time as a whole. The waking state of consciousness can be described as primarily the experience of succession with duration implicit in the continuous awareness of self-identity.

A final site in the experienced long duration of precontact and tribal-living Aborigines is, or was, to be found in the very nature of ritual. Archaic and aboriginal people envisaged events not as constituting a linear, progressive history but as so many creative repetitions of primordial archetypes or exemplary models. Thus an experience of long duration comes about through the archaic ideology of the punctilious repetition of rites and rituals required of group members and results in participants experiencing an effervescent feeling of indissoluble connection with the cosmos and with the cosmic rhythms. The result is a "sacred history," preserved and transmitted through myth and ritual reenactment, that reactualizes the great events that are believed to have taken place as the result of supernatural beings or mythical heroes. Continuing participation in the ceremony and the ritual gradually leads to an "assent to the disclosed terms of life" (Stanner 1979: 122), that is, to an acceptance of the liturgical, moral order encoded in the rites and ceremonies and to an awareness and appreciation of the Abiding Law. What is done in the ritual has been done before, so the experience of inner reality comes about through the ceaseless repetition of gestures initiated by others (Eliade [1954] 1991: xii–5). The prescribed gestures of the ritual reveal a primordial reality, this reality being a function of imitation of a celestial archetype, which, in the case of the Aborigine, is projected onto the enduring landscape.

Out of the past comes a cryptic determination for the here and now. Camp life is for witnessing, but this spatial openness is not extended in time. "Instead," Sansom (1980) concludes, "… there is *a temporal closure. It is the present not the past that is open to inspection* even though a version of the detailed past could be easily represented" (86, emphasis added). In this society, one comes to hear what everyone thinks about everyone else not through gossip about history but rather through commentaries and moral rhetoric on current action, on "what happenin'." Thus the establishment of consensus is both a form of closure and accomplished in the present. This is precisely what is meant by a natural experience of time, an experience that is at once patterned-cyclical and present-oriented.

Chapter 11

Equality Matching and Immediate-Participatory Time Consciousness

> I do not wish to create an impression of a social life
> without egoism, without vitality, without cross-purposes,
> or without conflict. Indeed, there is plenty of all, as
> there is of malice, enmity, bad faith, and violence. ...
> But this essential humanity exists, and runs its course,
> within a system whose first principle is the preservation
> of balance. And, arching over it all, is the logos of
> the Dreaming. ... Equilibrium ennobled is "abidingness."
> —W. E. H. Stanner, *White Man Got No Dreaming*

In this chapter the social relation equality matching (EM) is linked to the primordial, immediate, participatory kind of temporality described earlier by Whitehead (1929), James (1890), and Dōgen (1972) and in this chapter by Abram (1996) and key phenomenologists. It is proposed that the greater the extent to which members of a culture are involved in the *positive* experience of equality-matching social relationships, the greater their experience should be of immediate-participatory time consciousness. Members of indigenous, oral cultures, and the Aborigines to a high degree, because of their emphasis on attaining and maintaining a conditional equality in their social life, have organized their world in such a way that primacy is given to a participatory immediacy. Such equal social relations, it is proposed, involve efforts to attain, and maintain, an equivalence of thought, mood, and behavior, that is, a like-mindedness.

The Aborigines place great importance on the principle of equality. In totemism, the animal, plant, or natural force and the clan members who bear the totem's name are equal in belonging to the sacred world. As Durkheim insightfully observed, "Their relations are ... those of two beings who are on the same level and are of equal value." Totemic spirits are referred to in a way that denotes "a sentiment of equality. The totemic animal is called the friend or the elder brother of its human fellows," and members of the totemic species are seen as "kindly associates upon whose aid they think they can rely" ([1912] 1965: 162).

The Aborigines in their relations with each other place a profound importance on social equality, on making things "all-level," and to attaining, and maintaining, a congenial consensus in which everyone has an equal voice in developing the collective consensus. Throughout Australia, Aborigines tend to treat each other as equals, placing a premium on agreement, consensus, and like-mindedness. Great value is placed on shared experience. For Aboriginal countrymen, the source of all realized value is apt to be called "going through somethin longa somefellas." Those who go through significant events together are made "all same." Sansom explains the Aboriginal principle of consociate identity: "shared experience makes sharers [the] same; experience unshared signals, proves, and constitutes difference" (1980: 137). Non-participants lose, and are considered to have "thrown away," the opportunity for shared experience. Tribal-living Aborigines throughout Australia have no chiefs and headmen, with decision making being left to the collective wisdom of the elders and, in certain extraordinary circumstances, to "clever fellows" or sorcerers. As in communal sharing, the relationship between cultural meaning and social action can be explored by placing everyday social life in temporal context, in a way that is similar to that embodied in the notion of "cultural reproduction" (Myers 1986: 14; also see Bourdieu 1977; Sahlins 1981; Williams 1977).

Equality matching is based on social action that "has the capability to modify an existing state of balance in a designated set of social relations" (Sansom 1980: 195). When social events upset the social balance in a way that challenges the interests or even survival of the group, the Darwin fringe-dwelling Aborigines might say, "You gotta organize." The social demography of Aboriginal countrymen is based on the notion of balance. Sansom refers both to inter-mob balance and to intra-mob balance. In speaking of such balancing, countrymen might use the phrase "all level" and its counterpoint, "not level." To "level up" or "come level" refers to the leveling process. Two mobs might reestablish balance with the snatching of a potential wife of one mob by the other, with the celebrating captors saying, "This mob n ol Frank mob, wefella all level now." Such devices of settling intercommunity differences, and also of resolving interpersonal indebtedness, accomplish a state of equality matching. "To come clear" means to achieve balance in relationships that have a relevant history of consociation (Sansom 1980: 196–98).

Acceptance: Respect, Time, and Ritual

It is proposed that equality matching generalizes (but does not replace) Plutchik's ([1962] 1991) existential problem of Identity. Recall that acceptance and disgust/rejection are the primary emotions associated with this dimension. Thus we might expect the positive experience of EM to involve the emotion of acceptance or respect. Durkheim made just this observation: "Respect is the emotion which we experience when we feel this interior and wholly spiritual pressure operating upon us" ([1912] 1965: 236–37). And Stanner writes that the Aborigines' contact with the sacred leads to "a mood or spirit of 'assent' " (1979: 34). In the presence of the profane, the negation of the sacred, the emotion we experience is opposite of this acceptance or respect, namely, disgust and the dark side of the supernatural "forces" of the Dreaming.

The *tjuringa* of the Aranda was seen by Spencer and Gillen (1925) as sacred because it serves as the residence of a soul or spirit being. How the meaning of the *tjuringa* and other ritual objects is represented in the mind is not important. What is important, Durkheim claimed, is that "an object is sacred because it inspires, in one way or another, a collective sentiment of respect which removes it from profane touches." The sacredness of the *tjuringa* results "because of the collective sentiment of respect inspired by the totemic emblem carved upon its surface." Durkheim argued that society holds an empire over the consciousness of the societal member. This "empire," Durkheim maintained, "... is due much less to the physical supremacy of which it has the privilege than to the moral authority with which it is invested. If we yield to its orders, it is not merely because it is strong enough to triumph over our resistance; it is primarily because it is the object of a venerable respect" ([1912] 1965: 237).

Durkheim ([1912] 1965) associates his fundamental duality of the sacred and the profane with the primary emotions of acceptance ("respect," "reverence") and disgust. The "sacred : profane :: acceptance : disgust" association, as applied to the Aborigines' religion, is made in plain language:

> ... the whole religious life [of the Aborigines] gravitates about two contrary poles between which there is the same opposition as between the pure and the impure, the saint and the sacrilegious, the divine and the

diabolical. ... Of course, the sentiments inspired by the two are not identical: respect is one thing, disgust and horror another. (456)

The emotion of acceptance played an important role in Durkheim's research program for his new science of sociology, which endeavored to reconcile external, coercive society and individual volition. The solution, in the religious sociology of his mature work, was to see that the social order would be accepted because it was held to be sacred. The emotion of acceptance is closely linked to the ritual life. As Rappaport (1992: 6) explains, in the performance of a ritual the participants accept, and indicate to themselves and to others that they accept, the moral order encoded in the ritual. Thus the emotion "acceptance" expressed in ritual or ceremony is constitutive of the sacred, in the Aborigines' case, constitutive of the Dreaming as worldview and cosmology. The behavior specified by The Law thus provides for the Aborigine a positive identity. There is, in rituals, a celebration of the identification of the community and the larger ecological field, ensuring an appropriate flow of nourishment from the totemic landscape to the human inhabitants back to the local earth. The result is that the relationship between human society and the larger society of totemic beings is balanced and reciprocal.

From the symbiotic relationship of man and land expressed in ritual, there grows, *illud tempus*, a need to ratify, and socially signify, acceptance of this shared identity through the enactment of ritual and ceremony, in Cowan's terms, "so that a bridge could be built between the mundane and supramundane worlds" (1992: 28). In knowing more about the Dreaming, comprehension and acceptance of the state of the earth are uncovered, and this knowledge can then be passed on to tribal members who are "not privy to this esoteric information in a way that restores their faith in the cosmic order" (Cowan 1992: 39). Continuing participation in ceremony and the ritual gradually leads to an acceptance of the liturgical order encoded in the rites and ceremonies and to the stabilization of personal identity through awareness of, and inculcation in, the Abiding Law.

For the male Aborigine living in his traditional society, the means of gaining full status and adult identity is the initiation process. According to Munn, among the Walbiri and the Pitjantjatjara, what is transmitted in initiation is "not merely a particular kind of object and meaning content ... but also a particular form or mode of experiencing the world in which symbols of collectivity are constantly recharged with intimations of self" (1970: 157–58). Thus the self is identified with the external environment. The creativity and the autonomy of the person, according to Munn, belong to a "being split off from the individual and outside him, bound to the ancestral past" (1970: 159), which is the locus of free creativity (Myers 1972: 142). This kind of autonomy, Munn explains, is "reintegrated into the individual through identification with the external world—that is, through submission to the ancestral 'givens' of existence. Its attainment can never be separated from receptive submission to a given order of things" (1970: 159).

Munn carried out an analysis of myth and ritual of the Wawilak, which shows a circuit "from the subjective experience of an individual Ego through the objectification of collective codes (the myth and the sorcery codes) ... and back into individual experience *via* the social action of the rituals in which Ego is identified with the persons in the myth" (1969: 198). Myers (1972: 143) infers from Munn's analysis that the separation of experience from a subject through the objectifying codes is symbolic and occurs among the Murngin (Yolngu) in what Munn calls "symbolic space [which is] coded in Murngin thought as *time*: the events of the myth occur in an ancestral period distal from the individual" (1969: 199). By "time," Munn means that the "Reduction of the symbolic distance between an actor and the 'communalized' myth integrates the past (The Dreaming) within the present" (*ibid.*).

Equality and Consensus

Aborigines exert little power over things. There is little use for a large store of food that will not keep or a pile of spears too heavy to easily move about. Stanner (1979) describes Aboriginal society as one in which the primary virtues are generosity and fair dealing, as he writes:

> Nearly every social affair involving goods—food in the family, payments in marriage, inter-tribal exchanges—is heavily influenced by equalitarian notions; a notion of reciprocity as a moral obligation; a notion of generously equivalent return; and a surprisingly clear notion of fair dealing, or making things "level" as the blackfellow calls it in English. (40)

Of course, Aboriginal communities are characterized not only by harmony but also by discord. Canons of equality and fairness govern the occasional instances of armed conflict. The significant Aboriginal public fights are constrained by canons and conventions. Stanner (1968) reports:

> ... in the great public fights which ... were ritualized affairs, it would have been bad form to use a heavy killing spear against an opponent armed only with a light dueling spear. ... There was a distinct canon of equality at arms, a norm of sufficient—but just sufficient—retaliation, and a scale of equivalent injury. (48)

As a second example, one might find

> ... two men, both with a grievance each clasping the other with an arm and using the free arm to gouge the other's back with a knife; or in two unreconciled women giving each other whack for whack with clubs, but in strict rotation. (Stanner 1968: 48)

Rose provides a third example from the Yarralin:

> When Yarralin people are deeply grieved or deeply affronted, they rise up in anger, saying that only blood will satisfy them. The emotionally satisfying response to one's own loss is to inflict a loss. Balance is not only an abstract meta-rule; it is the content of emotional life expressed in daily exchange, politics, ceremonies, in life and in death. (1992: 223)

The lack of egoism in Aboriginal society is not, as Durkheim ([1912] 1965: 5–6) wrongly claimed, evidence of conformity of conduct. The unanimity attained in Aboriginal social interactions Durkheim saw as evidence that there is little individuality but massive conformity in Aboriginal society. There is, however, no evidence that this is the case. On the contrary, individual personalities are important in the creative social activity of attaining harmony, congeniality, and consensus among Aborigines, all of which can be attained without egoism. The individuality of Aborigines also is supported by the observation that Aborigines will not speak for each other and are apt to resent having another person speak for them. Further, a single Aborigine disagreeing with all other members of a group might have a decision delayed, or overturned, on his or her behalf. On the level of personality, Aborigines most certainly do not, as Durkheim implied, lack individuality, as Aboriginal communities are hardly bereft of real "characters." Aborigines are not at all self-promoting and do not wish to draw attention to themselves. Consistent with this, there is apt to be no ritualized talk in meeting others and then leaving their company. Both entrances and departures are unobtrusive and not marked by declarations of presence or impending absence (Liberman 1985: 27).

One manifestation of the normative order discouraging egotism and selfishness, as opposed to communal sharing, is feigning. Von Sturmer (1981: 17) sees feigning as an important part of public encounters that displays a desire not to present oneself too forcefully nor link oneself too closely with one's ideas. He writes that feigning is important in public encounters: "The people most vitally involved in an issue may feign disinterest, sitting with elaborate casualness; those whom one would expect to be the first to arrive in a public discussion may be the last to appear; the person most eager to dance may light up a cigarette just at the moment it seems his desire may be greatest" (*ibid.*).

Duncan (1986) sees embarrassment (*kunta* for the Pintupi), a mild form of shame, as an emotion that contributes to equality of status in Aboriginal society. Young Pintupi men rarely stand up to speak out of concern that they might make too much out of themselves. Similar comparisons also are seen in the conduct of older men, who typically begin speeches with self-deprecation such as: "I'm just going to tell you a little story," or "I don't expect you to listen to me." The implicit message is that a person should not be conceited, cheeky, or pushy and not think of himself or herself as being better than other people. In public ceremonies or on certain occasions, men avoid interrupting and contradicting one another (Myers 1979: 362). In Aboriginal society throughout Australia, direct questions and answers generally are avoided. Especially among women, there is a reluctance to speak out (Duncan 1986: 50–51). Aboriginal visitors coming to a different "country" will politely wait outside, waiting for those of the country coming to greet them and extend an invitation. Once invited in, they are apt to maintain a low profile and act in a circumspect way. In everyday social relations, Aborigines have an openness with respect to personal space yet stress privacy for their inner world.

The interactions of tribal-living Aborigines are not apt to involve the choice between different ideas and courses of action. There is little, if any, "either/or" and "if-then" reasoning involved. In fact, at least for the desert-living Aborigines studied by Liberman (1985), there is no word for "or" (but there is a silent equivalent). With only one idea under discussion, the process is one of moving toward a consensus. The organization of consensus, the matching of ideas across societal members, is likely in Aboriginal culture a protracted process. In this process, time is relatively unimportant. An Australian Senate Select Committee notes the following phenomena:

> For Aborigines, time is not of the same overriding importance as it is in virtually all non-Aboriginal affairs. Members of an Aboriginal community tend to talk at length around issues and so arrive at a consensus. In exchange between them and non-Aborigines the contrast in attitudes to time is of crucial importance. Failure to take due account of this contrast has contributed to poor communication of Aboriginal views. (Senate 1976: 31)

Liberman (1985: 20) correctly points out that the use of "time" in this passage, while not incorrect, glosses over the differences of social interactions under the heading "time." Liberman carried out a rigorous analysis of ordinary interactions and decision-making processes based on verbatim transcripts of conversations. Consensus in the absence of any formal leader or headman is produced through the practice of verbally formulating and acknowledging, and "thereby making publicly available—the developing accounts of the state of affairs which is emerging anonymously as a collaborative product" (Liberman 1985: 16). Individuals are proscribed from too strongly advocating their individual opinion; instead, points of view are more corporate than individualized. Meggitt (1962: 244) reports that an Aboriginal elder told him that for important matters, the members of the community should "all think in the same way with the same head, not different ways." Finding a consensus can be time consuming and difficult. As Liberman writes, "The process may take a great deal of time, but time is something Aboriginal people have in abundance" (1985: 16). In a meeting, Aborigines are careful not to speak at length; there is rather a checking with the assembly to renew one's (here, A's) license to speak, to verify that one is "speaking correctly":

A	All right?	(*Munta?*)
B	Yes.	(*Yuwa.*)
C	Yes.	(*Yuwa.*)
A	O.K.?	(*Palwa?*)
B	Yes.	(*Uwa.*)
C	Yes.	(*Uwa.*)
A	… like this, is this O.K.?	(… *alatji. Palyaku munta?*)
B	Yes.	(*Yu Wa.*)

Aborigines do not interfere with each other, give each other a lot of space, and do not pursue the details of matters about which they might not agree. Far from being uncivilized, Aboriginal social discourse is marked by humility, in which Aborigines make ideas and events available for public affirmations and the production of group cohesion, which is tantamount to a celebration of the collective life over that of its individual members (Liberman 1985: 153). If an Aborigine does not care to converse, then he or she will not speak. Often Aborigines will not speak when addressed directly (which can be highly disconcerting to non-Aboriginal persons). Their comments tend to be impersonal, not self-centered, instead seeking *emotional* identification with the group. Attention is centered not on the self, or even on other selves, but rather on the objective of the assembly. Consensus must be unanimous, attained without coercion, and sustained by congenial fellowship.

In tribal societies everyone is, to a much greater extent than in modern societies, constrained to think and act alike, and there is little economic specialization. Social solidarity is based on the fact that individuals resemble and like each other, as they believe in the same religious ideas and objects, share the same values, and in general feel the same collective sentiments. In this Durkheimian ([1893] 1933) "mechanical solidarity," consensus is due to the psychological similarity that results from sharing nearly identical beliefs and norms.

In contemporary Aboriginal life, knowledge of other persons is time-gapped as individuals of the same "mob" come and go, or are close or distant to "the hearth of a leader's wife" (Sansom 1980: 19). Holding to tribal customs, beliefs, and traditions leads to a societal consensus and a like-mindedness. Aborigines are predisposed to make decisions not individually but collectively. Consider the case of the Darwin fringe dwellers. In their informal interaction rituals, happenings come to be collectively represented, based on testimony and witnessing of the details and circumstances surrounding the events in question. Aborigines of the fringe mob keep track of all mob members, so contributions to the story about an event being socially constructed are based on place, on "watchin close up" and "watchin far." Discussions can eventually lead to a meeting of the camp leaders. "If all camp leaders come together, a unity of the whole mob is recognized" (Sansom 1980: 112). The time period spanning the occurrence of a significant event, a happening, and its resolution are recognized as times of trouble. From the moment it is recognized that some person or mob has a troublesome situation, those responsible for dealing with the trouble will work to "check up la all that detail," to "get all that detail right," and "to craft a story that can be approved because its narrative is 'straight'" (Sansom 1980: 128). Trouble is resolved in a public consensus, in which the "word" for ending the trouble is announced. Once a problematic situation or happening is fully resolved, constitutive rules hold that the incident, or the name of a dead person after "dead body business" is completed, should no longer be mentioned. At Wallaby Cross, the calling of a dead man's name evokes a rally for emergency. Such offending words, as Sansom explains, "symbolically undo the proscription of the past. ... When that [social] order is based on synthesis of words and happenings, the raising of problems from the past is an act of degradation in which the constituted reality of a mob is exploded" (Samson 1980: 159).

In this work of producing a Durkheimian "mechanical solidarity," equality matching and communal sharing are the involved kinds of social relationships. This is made clear by Sansom: "In the end, a story belongs to a whole set of people who as equal witnesses constitute the jurisdiction of a now fully fashioned story to the making of which they have each contributed assent" (1980: 131). Aborigines, Sansom found, on occasion compose both an "inside" word and a different word for the consumption of outsiders. Durkheim saw that those who are united in mechanical solidarity are united in likeness of belief. As Sansom explains, "In subscribing to their mob's word, the people of a mob declare their solidarity in that which 'you really can belieb' " (1980: 216).

In Aboriginal society, oscillatory cycles of reciprocal exchange also have high priority on the aesthetic level. This cultural emphasis takes the form of dancing and singing in synchrony and in the various other rhythms of formal rituals and everyday life. The vitality of Aboriginal social life that is displayed in their rituals and ceremonies is described by Liberman: "As the Aboriginal sit in subsection groups around their ceremonial fires, singing the stories of the Dreaming which are the spiritual and 'historical' foundations of their lives, the celebrants continue to chant in unison until after many hours—closer toward morning—they seem to be fused into one collective life" (1985: 6). Berndt and Berndt provide a similar description of Aboriginal ritual activity.

> The singing ... was ... vividly expressed, much energy, both emotional and physical being expended; the singers became part of the song, completely merging their individual identities in the rhythm of the wording. The voices were blended, with no lagging behind and no discordant note, giving the impression of one voice only, constantly varying in rhythm and tone. (1942–45, vol. 15(3): 242)

Durkheim ([1912] 1965: 242) recognized the importance of the bodily expression of such "social sentiments" for rituals, as it provided for the feeling of vitality and well-being that could be shared, as a collective representation, of an entire community. Commenting on the ritual and religious life of central-desert Aborigines as described in early ethnographies, he saw these social sentiments to be the origin of the collective life and the "communion of minds," which in turn resulted in mutual comfort. This communion of minds involved both mood and thought. Liberman (1985: 5–32) similarly identifies "congenial fellowship" as a basic organizing principle of Aboriginal life, an observation made by many, including the author, who have lived in Aboriginal families and communities. This congeniality is thus in evidence not only in formal rituals and ceremonies, but it also permeates the everyday life-world of the Aborigine.

The congeniality of fellowship among Aborigines—which includes a like-mindedness of both mood and thought—has been briefly suggested in a few ethnographic materials and reports (Myers 1986; Tonkinson 1978; Sansom 1980; Maddock 1972). Liberman (1985) has taken a further step, finding the principle of Aboriginal congenial fellowship plainly evident in ordinary discourse. A study of such conversations indicates that Aborigines expend *great effort* to share their fellowship. This is an important point, because just as great effort is made to create a conditional equality in social life, so also we have seen that great effort is required to fully realize an immediate-participatory time consciousness. Dōgen realized that his path to enlightenment was not along the linear time line but rather was compressed into a single, unexpected moment of maximum exertion. This also corresponds, based on personal communication, to the Aboriginal notion of Eternity, which is seen not as a permanent survival of the mind and personality following death but as a profound moment of shared wisdom.

The potential future, such as a proposed trip or event, is brought about, in Heidegger's terms "disclosed," only slowly and gradually, so that (1) a clear consensus has emerged and been made public, as assent, in conversation, and (2) the environment in which such a consensus has been attained is predicated on a friendly solidarity among the participants.

A Present Orientation

One obvious meaning of immediacy is its implications of a present orientation. Rose (1992: 206) writes that for the Yarralin people, the "now" as a temporal locus is differentiated from the past and the future on a number of bases, most of which are indicative of attempts to overcome the discontinuity of the before and the after. The we who are "here and now," living in a shared present, are seen as coming after the early-days people, who made the conditions of present existence possible in the first place. In relation to these earlier generations, the Yarralin refer to themselves collectively as the "behind mob," as those who come after the olden times of the Dreaming. The future, in contrast, is referred to as the "new mob," or as those who are "behind the we," and as those who come after the now. "It is our job," the Yarralin say, "to assure that those who come behind us are taught the Law and have a place and responsibility to take over from us" (Rose 1992: 206). "More profoundly," Rose adds, "although increasingly less successful, Yarralin people seek to assure that the behind mob will inherit what their own forbearers delivered: a world in balance and the knowledge of how to keep it so" (*ibid.*). Thus we see a link between the Aborigines' valuing "balance" and the immediacy of time consciousness. Equally obvious is what has been observed in many contexts, an assimilation of time to spatial relations, as the past is "behind," and the future is "ahead."

Aboriginal totemism objectifies the belief that all human and other beings are equal in that they share a common life essence derived from the ancestral spirits and creator beings. Rudder (1993: 31) explains that the Dreaming Events, which occurred at an "other" time, is simultaneously present through different physical manifestations in current time and through the present inner reality of the individual. At the ceremonial times, the presence of this, and other, impersonal forces can be felt and even seen by some (Berndt and Berndt 1948: 311). Such sightings *by* clever fellows as well as sightings *of* clever fellows who have died are not at all unusual phenomenal experiences for Aborigines, not only among tribal-living Aborigines but in detribalized and partially assimilated Aborigines in rural and urban settings. Such encounters, as was evident in the author's corpus of life-historical interviews with Aborigines, often involve receiving advice from the encountered clever fellow, which is apt to result in turning points of life as this advice is followed.

Hausfeld (1967) makes a theoretical comparison of Aboriginal and European value orientations. He proposes that Aborigines emphasize the present and the past, the nonmaterial, sociability, harmony with nature, and "being." Euro-Australians, in contrast, emphasize the future and the present, materialism, mastery over nature, and "doing." The "core culture" of modern, Western society is characterized, among many other features, by an emphasis on delayed gratification of needs and desires. This emphasis means that the individual is socialized to develop goals at various future locations for which he or she aspires and is willing to work (Holm 1974: 1). Settled, modern life makes planning for the future potentially rewarding and makes postponement of gratification a rational choice. Those with low socioeconomic status are not apt to share this middle-class to upper-class futural orientation. Many urban Aborigines live in slums. Corwin (1965) suggests that slum life is unpredictable and uncertain and focuses the attention of slum people on the present, or immediate problems that arise.

Aborigines might be thought to be oriented to the past to the extent that they stress their traditions. Few Aborigines today are living in a wholly traditional fashion, free of outside influences. Nevertheless, an appreciable number of Aborigines in the North and parts of the Centre are still oriented toward their ancient traditions. A study by Kearney and Fitzpatrick (1976) indicated that Aborigines with high level of acculturation had a more futuristic time orientation and experienced less group cohesion than did members of a less acculturated group. But there is no guarantee that contact with Westerners will inculcate in Aborigines an emphasis on planning and goal setting, because they do not feel welcome in the dominant,

Euro-Australian culture and are apt to systematically experience prejudice, discrimination, and racism. Holm (1974), in this connection, carried out an empirical study of time perspective in three Northern Territory Aboriginal communities. He studied 120 children, ages nine to eighteen. He found that Aboriginal boys who have been raised in a traditional way, being initiated, in a Northern Territory community studied were no more past oriented (and were, surprisingly, more future oriented) than were boys raised in two other communities, one with medium and one with high contact with Western society. The finding that contact with Western society did not result in a high level of future orientation was unexpected. A possible explanation was that these Aboriginal youth learned from this contact the harsh reality of racism, prejudice, inequality, and restrictions of their possible future opportunities, which in turn might well have resulted in lowered self-esteem and a limitation of future time orientation.

In summarizing his analysis of the determination of social, political, and religious institutions with economic geography and the totemic landscape, T. G. H. Strehlow (1970) describes the Central Australian Aborigines as living in a land where supernatural beings are revered and honored by their human incarnations. These supernatural beings were seen not as living in Heaven or in the Sky but rather as manifesting themselves "in the mountains, the springs, the sandhills, and the plains, religious acts had an *immediate personal intimacy* rarely, if ever, equalled in other religious systems" (133, emphasis added).

For the sedentary person, his or her environment is an objective reality that has been created out of human imagination and constitutes an objective reality—whether a castle, a village square, a tombstone, or a house. For the tribal Aborigine, the physical environment is rather a projection of the archetypes governing his existence, real or imagined. The Aborigines' traditional orientation to space and to the land is reflected in social behavior. As Cowan notes, "Since his space is not constricted by numbers nor by the idea of time, he is able to adopt an entirely different posture, both at the intellectual and the physical levels" (1992: 90). Whether in action or in repose, "the Aborigine is adapted to spaciousness" (Cowan 1992: 90). Aborigines rarely sit close to one another; in fact, when they do come together to talk, they invariably sit about with a great deal of space between them. It is as though they wish to preserve a feeling of "distance" in keeping with their understanding of its limitlessness. Thus the embodied actions of the Aborigine carry with it a notion of an eternal present and at the same time a notion of unlimited space. The Dreaming, Myers writes, "possesses no single or finite significance. It represents instead a projection into symbolic space of various social processes" (1986: 47). Here, as elsewhere, we find the Aborigine shifting time into a spatial habitus. An emphasis on the present, then, results from immersion in the sensory world continuing to manifest itself in the immediate and finding its expression not in a Heideggerian temporal stretch but rather in spatial extension.

For the tribal-living Aborigine, the prospect of death leads to talk not about time but about identity with the land. Cowan cites that Aborigine Bill Neidjie, responding to a question about time, quickly shifts his focus to the existence of the physical world in the present and his involvement—bodily, social, and spatial—with the land and with nature:

> When the law started? I don't know how many thousands of years. Europeans say 40 000 years, but I reckon myself probably was more because … it is sacred. Rock stays, earth stays. I die, I put my bones in cave or earth. Soon my bones become earth. … My spirit has gone back to my country. We always use what we got … old people and me. I look after my country, now lily coming back. Lily, nuts, birds, fish … whole lot coming back. We got to look after everything, can't waste anything. Old people tell me, "You got to keep law." "What for," I said, "No matter we die but the law … you got to keep it. You can't break law. Law must stay." (Cowan 1992: 97)

The limited technology of a hunting-and-gathering society such as the Australian Aborigines' provides little chance for controlling the outcomes of future events, so societal members tend to live from day to day. Overall, Aborigines share with a wide variety of nonmodern cultural groups a present orientation. Attention is given to the present and its practical, daily concerns. Aborigines are apt to say that they "take notice" of their country. Activity and nonactivity are not distinguished in terms of value. The present, for traditional Aborigines, is not evaluated in terms of its contribution to a future, worthy end but is good, or not good, in and of itself. Aboriginal children throughout Australia now typically attend school but are not likely to have a strong orientation to schooling as preparation for the future; rather, a day of school is apt to be viewed as an end in itself (S. Harris 1984: 28). This kind of time consciousness—with different concepts of time, different attitudes of time, and different habits of making use of time—is likely to be at odds with the expectations and demands of a formal, mass educational system. Misunderstanding can result (S. Harris 1984: 1). Both Aboriginal and lower-class non-Aboriginal children appear to have a limited time perspective, yet their schools are future oriented, creating a lack of temporal congruence between the school and the child (Holm 1974: 6). Even in

Aboriginal hunting-and-gathering activities, no provision was traditionally made for the future, this having been left to various increase ceremonies. The present orientation of Aboriginal children is not what the formal educational system wants, for this system is intended to motivate children to work toward very long-term goals, eventually toward getting a job at the end of schooling and then actively participating in civic culture and the cash economy. Sommerlad writes, "The first value which education seeks to change is that of time" (1976: 52). And Gallacher sees that the emphasis on education is being placed on future orientation: "... to introduce them to our time values poses a problem of considerable magnitude, and regrettable though it may be ... this is one of the tasks which education must sympathetically concern itself" (1969: 100, cited in Sommerlad 1976: 52).

The assimilation of Aborigines was adopted as governmental policy in 1937, at a conference of state and federal officials convened by the Australian federal government. The assimilation policy was directed toward the inculcation of future-oriented goals, according to which Aborigines were to gradually learn to live in houses, work at paid jobs for a "training allowance," and seek education for socioeconomic upward mobility (Lovejoy 1975). Stanner has condemned a policy that Aborigines are to be treated as "individuals" and not as "groups." "No policy or law," he writes, "can transform the aborigine from which he is in this region—a social person tied to others by a dozen ties which are his life—into an abstract 'individual' in order to make the facts fit a policy" (1958: 100).

Even among Aborigines whose culture is highly eroded, the Aboriginal custom of sharing property as needed has been retained. As Beckett (1965) notes, even if ceremonies and rules have been forgotten, there is still a loosely defined obligation to share food, clothing, money, and more with kinfolk in need. The essential aspect of Aboriginal sociality, relatedness, is maintained and further developed. "Un-Aboriginal" forms of behavior showing exclusiveness, superiority, and being "flash" are disapproved and derisively called efforts to ape white people. Aborigines committed to their own culture are said to "identify," will not "knock back," reasonable requests from kin and other community members and will work to promote "Aboriginality." Consider the testimony of Ken Hampton, an Aborigine who was institutionalized, at Mulgoa, New South Wales, at age three, living with other members of what became an underprivileged family, largely confined to the cellars of their "home," and experiencing deprivation and oppression. He was later transferred to St. Francis House in Adelaide. When "released" as an adult, he was able to "learn to share with your wife and with your children, and you have to learn to communicate with other people"

(Mattingley and Hampton 1988: 139). Hampton's experience of a protective institution thus did not destroy him but did result in what must be called "cognitive assimilation," as he developed a Western consciousness of time. Hampton is quoted as saying, "I know I still act institutionalized. There's been a certain amount of regimentation in the way I've brought up my children. For instance, I've been very strict and uncompromising about time, what time they've got to be home" (Mattingley and Hampton 1988: 139).

Aborigines tend not to be "goal setters"; goals have to do with an imagined and a valued future state and require consciousness of the length of time between setting and achieving of goals (Banks 1983: 223). Their lack of future-value orientation has confounded and frustrated a protracted effort by the Australian government to force assimilation on Aborigines. The "prediction" that contact with the dominant, European society will result in Aborigines developing a future-value orientation has not been verified.

IDENTITY, CHANGE, AND IMMEDIACY

Writing of the western-desert Pintupi, Myers states: "*The Pintupi are dominated by immediacy*" (1986: 17, emphasis added). He adds, "Nothing seems settled unconditionally," then provides an illustration: "Thus, a man who deeply desires that a particular girl be married to him could, through intimidation, force her relatives to break a promise of bestowal to another" (1986: 17). Myers attributes Aboriginal "immediacy" to a context dependence on the part of Aborigines. Maintaining one's relationships with others is in Aboriginal society an end in itself. On the level of individual action, Myers writes: "I believe that the immediacy of *current* relations so dominates Pintupi social life that the production of an enduring structure that transcends the immediate and present is a cultural problem for the Pintupi, for other Aboriginal people, and for many other cultures as well" (1986: 17, emphasis in original).

We have seen that the mythological construction of the Aborigines, the Dreaming, symbolically links everyday life to the enduring features of the totemic landscape, providing what Myers calls an "ontological order." This Order, the Dreaming, "provides a means of surmounting the constraints imposed by the need to sustain immediacy" (Myers 1986: 17). The transcendence of immediacy is an accomplishment of shamanistic mentation. There would appear to be little difference between the Aboriginal clever fellow's apprehension of the separate reality of the Dreaming and mystical experiences in other cultures and religions (Kalweit 1988: 241).

This Aldous Huxley (1946) termed the *philosophia perennis*. Eliade (1951) is one scholar who sees such a linkage between shamanism and Buddhist mysticism, as he writes: "Shamanism represents the most credible mystical experience of the religious world of primitives and, within this archaic world, fulfills the same world as does mysticism in the official faith of the great historical religions from Buddhism down to Christianity" (1951: 96, cited in Kalweit 1988: 241). Thus there is some level of continuity in temporal experience between the peak experience of the Aboriginal man of the highest degree and that of Dōgen's enlightenment. Both share an immediate-participatory time consciousness. On this point, Kalweit writes of the expanded outlook of such an experience: "The realization of unity— brings about a sense of sacredness, of transcending time and space, and the absolute certainty that present, past, and future are artificial concepts" (1988: 243). This temporality is clearly of the immediate-participatory kind, as Kalweit adds, "At the same time there is a feeling of being totally submerged in the moment and guided by a higher power from which grows the conviction that a supersensory source of energy exists" (*ibid.*).

A Norm of Reciprocity

One vital aspect of equality matching is that of exchange and reciprocity. Malinowski ([1922] 1961), in his ethnographic work with the Trobriand Islanders, described a system of the exchange of services, goods, and obligations, which he distinguished from nonmaterial, symbolic exchange. In his classical study of gift, Mauss ([1925] 1954) elaborated on Malinowski's theory of exchange by distinguishing between material exchange articulated in accordance with codes of socioeconomic behavior and symbolic exchange linked to religious practice. In the case of the Aborigines, we find both material and symbolic exchange, both of which create a sense of community and— through exchange and cultural borrowing beyond tribal boundaries—the development of a Pan-Aboriginal worldview and shared notions of temporality. In a gift exchange, people are bound to each other through the agency of things and in which the things become imbued with notions of self. Objects for exchange "are symbols of the social relationships they define, because they condense within themselves the structure of those relationships. They are, in effect, icons or expressive symbols of them" (Munn 1970: 141). Aborigines differ from Westerners in that they place great emphasis on symbolic-religious exchange and little emphasis on objects that can be regarded as pure commodities, as in modern societies with competitive market economies.

There is evidence that Aborigines enjoy these dynamic aspects of their religious life and seek to develop them further. There is in ritual life a concentration of the mind that represents a celebration of the moment. Durkheim saw this intensification of the moment as the primary difference in the experience of the profane, everyday world and the sacred, inner world: "*By concentrating itself almost entirely in certain determined moments, the collective life has been able to attain its greatest intensity*" ([1912] 1965: 251, emphasis added). The sentiments so aroused attach themselves, as Durkheim realized, to the symbols that represent them, such as emblems of totems.

Because the Aborigines keep no records of religious history, anthropologists have had to confine themselves to a synchronic approach and look at sociocultural structures *as if* they existed in a timeless present. Kolig (1981) suggests that these same anthropologists have made a virtue out of necessity, concluding that the cultures they study are "ahistorical," "prehistorical," and "timeless"—having no history, no change, and no development. Radcliffe-Brown (1952), in light of this perspective, recommended that anthropology dispense with "pseudo-historical" explanations, rather carrying out synchronic comparisons. Exponents of this view, including Eliade ([1954] 1991) and Lévi-Strauss (1965), have argued that "primitive" people do not possess a linear conceptualization of the passage of time and so do not acknowledge the prospect of change and development in their societies. Lévi-Strauss (1966) distinguishes between "cold" societies that do not see change and development in their cultural forms and social structures, such as the Aboriginal societies, and "hot" societies, progressive and modern, that value change and development. Rosaldo (1980; also see Goody 1977) has provided an effective critique against this view that there exist "cold," "timeless," "primitive" societies, arguing that this characterization reflects not their basic nature of "primitive" society but a methodological and ideological bias in anthropological research.

Aboriginal political life, a struggle against all odds, does not fit this stereotype. And Aboriginal religious practices do not fit this stereotype; it is not at all timeless and unchanging but is rather dynamically developing in a historical context. In this process, exchange is a vital ingredient. Tonkinson, in his study of the Mardudjara Aborigines, found four sources of change and innovation in Aboriginal religion: "the diffusion of new rituals, songlines, and objects ...; "ritual innovation at the local level"; "discoveries of sacred objects"; and "the exploitation of myth's inherent flexibility" (1978: 134). With new rituals, Tonkinson notes the big meetings, where groups come together "in emulation of the Dreamtime beings who instituted and exchanged rituals and objects on many occasions when their paths crossed" (ibid.). In these meetings, religious knowledge is advanced, and men of adventurous spirits are stimulated to undertake Dreaming journeys into new areas. Rituals are owned by particular groups but may be exchanged with other groups, which can later modify them and thereby also claim ownership.

Reciprocity, an essential feature of Aboriginal culture, has by no means been abandoned among Aborigines deprived of their lands and living in the fringes of towns and cities. For example, Sansom (1980: 247) describes a practice among the fringe dwellers of Darwin, according to which several married couples pool their resources to establish a single hearth, with the women sharing the work of food preparation. A principle of symmetrical exchange prevails, so each married woman (with the right to her own hearth) contributes a fair share of the rations. This pooling of rations is kept "level up" by debt relationships, with those possessing cash helping those who cannot contribute. Rations also can be exchanged for services in complex ways. A person not contributing but "takin tucker" will quickly become known as a person engaged in "wrong kitchen business." Sansom (1980: 61) provides as a second example the way in which the consumption of alcohol is carried out. The fringe dwellers have made of "grog" [beer] "the commodity of immediate reciprocities," making of grogging sessions "a jointly experienced progression in which people 'go through' the stages of inebriation together and more or less in step so that co-drinkers remain 'all level'" (*ibid.*). In order to avoid misfortune and mortality in such sessions, Sansom found in these Aborigines a twofold strategy—"authoritative abstention" and "equal surrender." Drinking can take place under the supervision and protection of a credit-worthy leader, a "Masterful Man," who takes full responsibility for any untoward happenings that might be the result of a drinking session, who might be appointed to ensure that the rules of cup passing are followed, and who is kept in beer as payment for his work. By equal surrender, Sansom refers to a set of prophylactic conventions that govern behavior, say, around a flagon of (usually watered) port, in which there is an effort to see that each person gets a fair share, or "stays level."

The Phenomenology of Immediacy and Participation

In this chapter, and in this book, much attention has been given to the notion of a totemic landscape as a focus of Aboriginal spirituality. It should not be overlooked that a totemic landscape is a paleosymbolic image linked to the real landscape, to nature itself. Abram (1996) persuasively argues that what members of indigenous and oral cultures view with the greatest awe and respect is nature itself—the plants, animals, forests, mountains, and winds. It is the concern of the shaman, in Aboriginal culture the "Clever Fellow," to slip out of the perceptual boundaries that demarcate his or her culture in order to make contact with, and learn from, the other powers of the land, through a heightened receptivity to "… the meaningful solicitation—songs, cries, gestures, of the larger, more than human field" (Abram 1996: 9). This attunement to the environing nature of native cultures, Abram explains, "is linked to a more primordial, *participatory* mode of perception. … [P]erception is *always* participatory" (1996: 27, emphasis added; 276, emphasis in original). This participatory mode of information processing has been shown by phenomenologists such as Husserl ([1924] 1960) and Merleau-Ponty (1962) to turn toward the things themselves, the world itself, "as it is experienced in its felt *immediacy* (Abram 1996: 35, emphasis added). To this participatory immediacy we find the world as it is directly experienced. This intersubjective "life-world" (*Lebenswelt*) is, as Husserl explained, the world of our immediately lived experience. Husserl suggested that the earth, our "primitive home," lies at the heart of our notions of time and space. Husserl was limited by his lingering assumption of a self-sustaining, disembodied, transcendental ego. Merleau-Ponty rightly rejected this notion, recognizing that the living, attentive body, the "body-subject," makes it possible for the experiencing self to enter into relations with other presences. He realized that all of the creativity of the human intellect must be understood as an elaboration of the most immediate level of sensory perception. It is through our bodies, key structures in the phenomenal field, that we are able to perceive other bodies, recognizing them as other centers of experience, other subjectivities. Thus the phenomenal field is not a solitary ego, as Descartes believed, but a collective landscape, constituted by other experiencing subjects as well as by the self.

The best single term we could use to characterize the events of perception as disclosed by phenomenological attention is *participation*, used by anthropologist Lucien Lévy-Bruhl ([1910] 1985) to characterize the animistic logic of indigenous, oral people. For such people, including Australia's Aborigines, mountains, plants, and animals are felt to participate in one another's existence, influencing each other and being influenced in turn. For Lévy-Bruhl, participation was a perceived relationship between diverse phenomena. But Merleau-Ponty suggested that participation is a defining attribute of perception itself. Perception, phenomenologically considered, as the concerted activities of all of the body's senses acting together, is inherently participation. As Abram explains, "perception always involves at its most intimate level the experience of an active interplay, or coupling, between the perceiving body and that which it perceives. Prior to our verbal reflections, at the level of our spontaneous, sensorial engagement with the world around us, we are *all* animists" (1996: 57, emphasis in original). When the immediate-participatory mode of information processing is given emphasis, as it is in oral and indigenous cultures, the entire world seems to come awake and to speak. This incarnate, sensorial dimension of experience, Abram notes, "brings with it a recuperation of the living landscape in which we are corporeally embedded" (1996: 65). The bearer of an oral culture, then, sees plainly that his or her sensory perception is simply "our part of a vast, interpenetrating webwork of perceptions and sensations borne by countless other bodies" (Abram 1996: 65). This intertwined web of experience is the "life-world" of Husserl, the biosphere in which we are embedded as it is experience and lived from within by the intelligent body, the human living in the experienced world. From this perspective, we can see why Aborigines are so mindful and respectful of the watchful land itself. The Aborigines, along with other long-established indigenous cultures, display a remarkable solidarity with the lands that they inhabit, along with a respect, even a reverence, for the other species that also inhabit the land. Abram, from this perspective, sees the initiatory "walkabout" undertaken by Aboriginal Australians as "an act whereby oral peoples turn toward the more-than-human earth for the teachings that must vitalize and sustain the human community" (1996: 116). He adds: "In indigenous, oral cultures, nature itself is articulate; it speaks. The human voice in an oral culture is always to some extent participant with the voices of wolves, wind, and waves—participant, that is, with the encompassing discourse of an animate earth." No element of the landscape is definitely void of expressive resonance and power; any movement may be a gesture, any sound may be a voice, a meaningful utterance.

Social equality is a social universal, an aspect of social interactions in the everyday world of people in every society. In work, equality matching exists insofar as each person does the same thing in each phase of the work, either by working in synchrony or by taking turns. In decision making, EM means that everyone has one vote, or an equal say, or an assent. Equality matching is institutionalized in democratic governmental forms. The members of informal social groups usually are equal-status peers. As for social identity, the individual person is a separate but co-equal peer, on a par with one's fellows. If this form of social relation is insisted upon, then misfortunes might be, at least roughly, equally shared, so "things even out in the long run." On the level of natural selection mechanisms, there can be a "tit-for-tat," in-balanced, in-kind reciprocity, with appropriate temporal delay. In game theory, Axelrod (1984) has found that tit for tat—a strategy that cooperates on the first move and thereafter mirrors the other player's previous move—received the highest average score, implying that people act in their own best interest if they practice reciprocity. Trivers (1971) shows that "reciprocal altruism" has evolved through natural selection, existing among cleaner fish, dolphins, baboons, and humans. The principle of "reciprocity" solves the Hobbesian problem of how it is that individuals, acting in their self-interest and in the absence of a coercive authority, are nonetheless able to produce social order (Ellis 1971). Gouldner (1960: 173) saw reciprocity contributing to social order, because a debtor is morally bound not to harm a creditor, while a creditor will not wish to forgo the opportunity of being repaid. Blau's principle of "social exchange" holds that "people often do favors for their associates, and by doing so they obligate them to return favors" (1964: 314). The social psychology of reciprocity also has been used to explain why nation-states cooperate despite the absence of a coercive central authority (Larson 1988). According to the norm and logic of reciprocity, nation-states will be predisposed to act with reciprocal restraint on arms or trade barriers out of self-interest, thereby avoiding arms races, military conflicts, and trade wars.

Chapter 12

Authority Ranking and Episodic-Futural Time Consciousness

The two primary emotions associated with Plutchik's ([1962] 1991) life problem of "hierarchy" are anger and fear. Anger expresses itself behaviorally as a moving toward obstacles and goals in an effort to push them aside; fear is a defensive behavior, an effort to retreat, to move away from threat and danger, which can take the form of being incapacitated by a deficiency of social power (Barbalet 1998: 149–69). There is no conceptual difference between the concept "hierarchy" in Plutchik, MacLean (1973), and Scheler (1926) and the concept "authority ranking" in Fiske (1991).

AUTHORITY RANKING IN ABORIGINAL CULTURE

> [Aborigines] ... do not fight over land. There are no wars or invasions to seize territory. They do not enslave each other. There is no master-servant relation. There is no class division. There is no property or income. The result is a homeostasis, far reaching and stable.
> —W. E. H. Stanner, *White Man Got No Dreaming*

In this chapter, we will first briefly examine the Aborigines, whose societies have historically existed without politics, headmen, or clearly defined leaders, but in which hierarchy nonetheless exists, an important point insofar as hierarchy is considered a universal condition of human society. We have described Aborigines as having an immediate-participatory time consciousness, which means first of all a temporality focused on the present, on the sensed immediacy. No claim is being made that the traditional Aborigines are concerned *only* with the past and the present and not at all with the future. One of the most misleading stereotypes about Aborigines is that because they have traditionally lived a hand-to-mouth existence, one day at a time, they are thoughtless, not mindful, and improvident. Aborigines have been further stereotyped as being obsessed with the past and never bothering to plan for the future. That this characterizing is misleading, at best, can be easily seen in the forward planning that Aborigines carry out for social ceremonies such as weddings, ceremonies that must be carefully planned in a manner consistent with kinship rules, and in the planning of hunts and journeys (Berndt and Berndt 1978: 47–82). Strategies for coordinated action must be planned in advance of an actual hunt. For both hunting and gathering, the movements of Aborigines over a large area are motivated by a need to find the best sources of food and water over their various, overlapping "seasons."

Tribal-living Aborigines had available to them a plenitude of "instant foods"—fruits, wild honey, thin "wild potatoes" that can be eaten raw, wild onions and other kinds of roots, shellfish, and highly prized witchetty grubs. Many foodstuffs required cooking or some other treatment before eating. Future planning was, of course, necessary to find the product at its source, by collecting, digging, or killing. For men, this meant hunting expeditions, and for women, it meant field trips in which foods were gathered and, when opportunity presented itself, edible animals caught. Planning for family meals involves planning a sort of itinerary. There also must be preparation for ceremonies and rituals. In order to carry out significant and complex rituals, negotiations must be carried out, messengers must be dispatched to request the presence of scattered groups, and the timing of events within the ceremony itself is likely to be carefully coordinated. Actions within the ceremony must take place when signaled not by clocks but by events or happenings in the ceremony itself. For the Pintupi, "holding a country" (*kanyininpa ngurra*) can make an individual responsible for organizing significant events. Such religious leadership, usually held collectively by the elders, is one of the most important criteria in a community's collective prestige. Writing about the Aborigines of the Fitzroy River area of the Kimberleys, Kolig claims that a community lacking a religious elite will never rise above mediocrity. Members of this elite are not to act alone, as "an individual acting alone is considered helpless and worthless" (Kolig 1981: 125). The mere presence of religious leaders is not enough; instead, they must be willing to actively promote and stimulate religious involvement. Of course, the location of a community is itself important to its religious prestige, because its prestige resides in the earth itself. This status of place is not entirely fixed from the age of the Dreaming, however, because a locale can be imbued with sacredness through human ritual activity. The cache of sacred objects belonging to a community also can add to the community's prestige.

Perhaps the best example of long-term planning in Aboriginal society is the system of promised marriages, where a girl might be betrothed when only a baby, or even before birth. Children are prepared for sexual relationships in general, and for marriage in particular, from the time they are babies. This is done through formal instruction but also through teasing and conventionalized joking. In arranged marriages, a high value is placed on long-term economic security. Getting married is part of a wider constellation of actions and beliefs, part of the system of social planning for the future, and for ensuring continuity in social relations. Thus there exists a need for planning, even in tribal Aboriginal societies.

Collective Decision Making

Liberman (1985) and many others have described traditional Australian Aboriginal decision making as collective and nonhierarchical. In fully traditional Aboriginal societies there exists no formal leadership, and political life is remarkably egalitarian. Not only is their no chief or headman, but individuals are proscribed for advocating their own positions too strongly. The Aborigine point of view in any gathering is corporate rather than individualistic. Decisions are made by the group as a whole and, as we have seen, require the verbal assent of nearly every member present at a meeting, an event, or a gathering. When discussions become problematic, a central-desert Aborigine is apt to offer a comment such as, "Let's speak with one voice, harmoniously" (*wangka kutjungka walykumunu*) (Liberman 1985: 15–16).

Consistent with their egalitarian decision-making processes, Aborigines are conspicuously reluctant to exercise temporal priority (e.g., to be the first to speak on a topic), or to exert individualistic leadership. While important decisions will likely be made by the elders, often at the request of younger persons who lack the elders' knowledge of their tribal Law and of life in general, such decisions are nonetheless consensual and do not require what Collins (1990) refers to as "power rituals."

Up until this point we have endeavored to show the ways in which Aborigines' culture gives emphasis to and expresses two social relations, communal sharing and equality matching. It is proposed that traditional Aboriginal culture, by giving emphasis to CS and EM as the predominant forms of social relations, contributes to a time consciousness that emphasizes patterned-cyclical and immediate-participatory time. Moreover, insofar as the unity of CS and EM is constitutive of hedonic community, and the unity of patterned-cyclical and immediate-participatory time constitutes natural time, it is further proposed that involvement in a hedonic community contributes to a natural experience of time.

But this analysis leaves the relation of Aboriginal cosmology to temporality unexamined from a sociological perspective. This problem can be approached with the aid of Myers's analysis of the Pintupi, as he proposes, "The Pintupi appropriation of temporality is guided ... by the two basic problems of Pintupi social life, the tension between relatedness and differentiation and that between hierarchy and equality" (1986: 70). The related-different distinction corresponds, at least roughly, to the problem of communal sharing: when Pintupi people are making an effort to sustain close relations with family, hunting band, neighborhood, kin, clan, and tribe, their goal is to develop a publicly displayed and congenial consensus. A similar social process also has been observed among the fringe dwellers of Darwin (Sansom 1980). The consensus, the "word," is a gestalt synthesis of conversational details. Aboriginal thinking is correlative, and as Grey (1975) emphasizes, it maintains a gestalt summarization that is constantly updated. This makes possible a like-mindedness of thought and mood. We have seen, Durkheim's opinion-at-a-great-distance notwithstanding, that Aborigines are individualistic in their personalities and value autonomy in themselves and others as well. Reaching and maintaining a congenial consensus is an experience of duration and requires skillful talk and expression; success cannot be taken for granted, as the level of cultural and community disorganization of Aborigines can be such that the result is not a universal consensus by a division into hostile factions, or to the out-migration of those who wish to be supported only by outstations or other mobs, or to the rural Aboriginal community, or to the relative anonymity of suburban life.

Even in the midst of such disrupted communal life, there is, at least on the level of ordinary conversation, a premium placed on equality matching, as described earlier. The negative pole of equality matching is mismatching. Division of opinion emerges, and the interests and worldviews of community members no longer represent a consensus but seek an individually designed attitude toward the world. Social differentiation is the precondition for the emergence of social hierarchy and community leadership. When Aborigines are no longer living in the old tribal way, hierarchy and leadership *do* develop in their communities. The Aborigines living a marginal life in the camps and fringe settlements of Darwin must, Sansom writes, "... deal consistently with emergent states and realizations and with futures that are indeterminate because not meaningfully contained in any projections of likely career choice through predetermined hierarchies of status and rank" (1980: 190). Rather than following a prefigured course of action (e.g., going to school, then pursuing an occupational career), they instead proceed contingently from one state of affairs to the next, which is always unpredictable. They must be prepared, for instance, to defend themselves and their camps from attacks by "bodgies" they do not know to "rebels" they do know (perhaps as "mates" in working-class urban bars). They also must walk in groups in Darwin, with their company including a sufficient number of Fighting Men. Thus their leadership is oriented not to controlling the future but rather to coping with events that they can neither predict nor control.

The Authority of the Abiding Law

How are the tensions of these dualities, related-agreeing/different and matched/mismatched, to be released and resolved? If Fiske (1991) is correct in his claim that there are four elementary forms of human relations, then this resolution must involve at least one more relation, which is proposed—following Myers, to be authority ranking/hierarchy. This authority, for the Aborigine, is the Cosmological Order, the Abiding Law. As Myers (1986) puts it, "As an order outside of human subjectivity which is morally imperative, the Dreaming can also be understood in relationship to the problem of hierarchy in a small-scale society where egalitarian relations are valued, and to the organization of a regional system of sociality" (70). What Western people would understand by "authority" and "autonomy" are considered by the Pintupi to be "outside," projected outward into the Dreaming and onto the landscape, where they exist as artifacts. The landscape frames the individual's body; further, group identity is anchored in the land (Mol 1982: 6). The Dreaming manifests itself in text—in the dances, songs, stories, ceremonies, rites, and rituals of spiritual life. "Dreaming" refers to specific texts such as specific stories and the creative epoch of which the stories are a part. For example, "[a] narrative of the traveling ancestor who killed the dancing women becomes, then, the Monitor Lizard Dreaming" (*ibid.*). Thus there exist two levels of attention, one to the immediate participatory (which are *yuti*, visible or perceivable by some other sensory channel) and the other to the level of reality of the Dreaming (*tju-kurrpa*). The apprehension of spirits, creatures, and landscape features, then, is of the invisible. An effort to see the deeper meaning of what is immediately present, yet transcendental, is an experience of primordial temporality. An event, such as the sudden appearance of a "Clever Fellow," can be *yuti* even if such an apparition is not really seen but is, in principle, witnessable (Mol 1982: 49). The Dreaming, then, "constitutes the ground or foundation of the visible, present-day world.... The Dreaming provides an explanation for the existence of everything in the world" (Myers 1986: 70)—from persons, to geographical features of the landscape, to anthills. The Pintupi say, *tjukurrtjanu, mularrarrngu*—"from the dreaming, it becomes real," which means a passage from one plane of being to another. A similar duality can be found in the Yarralin, who see the human being participating in two kinds of life. *Manngyin* is the unique, the changing, and the irreplaceable. *Yimaruk* refers to the continuity of life. *Manngyin* never returns, while *Yimaruk* can never be lost. *Yimaruk* binds the societal member to the Dreaming in the sense that as a spirit that has been created, it cannot be washed out or destroyed. This spirit can change from one species to another. Rose explains that "cutting across boundaries of space and time, it ties us into the whole process of life: a process which, barring disaster, is everlasting" (1992: 215).

Knowledge, Authority, and Temporality

The Dreaming has a transcendental character. Its existence as a pattern of stories allows it to transcend the immediate present. That the Dreaming has a transcendental character does not mean that authority resides in the Law alone. Authority in tribal Aboriginal society belongs to those men who have knowledge of the Law. Religious knowledge is not common knowledge, rather, being carefully controlled. Keen (1994), in a 1970s' study of the Yolngu men of northeast Arnhem Land, examines the framing of religious forms and control of the dissemination of knowledge in three contexts: among social networks centered on patrifilial group identity; in age and gender relations; and in relations with people of a wider region, not excluding the use of more nearly universal symbols (e.g., Christ on the Cross).

Yolngu control of knowledge was structured according to their concept of *marnggi*, which means "know" or "knowledge" and also conveys a sense of "can" or "be able." Thus knowledge is not separable conceptually from power. Knowledge, for instance of stories and, dances, comes about from experience but also depends on age group, sex, and cultural group. Thus, for example, members of the same moiety might share negotiated religious forms but might be differentiated by the details of forms created in religious ceremonies and by the person's total complement of *sacra* (myths, songs, designs, objects, and ceremonies). Keen (1994: 2–3) described differences in the Yolngu interpretation of ancestral events, the meaning of place and country, and the meaning of ceremonial events. He described the interplay between, on the one hand, the definitions of *sacra*, connecting and differentiating group memberships, and, on the other hand, religious secrecy separating males and females as well as the young and the old.

Aboriginal society is largely egalitarian, and decision making and planning are carried out in a nonegoistic, consensual manner. At the same time, however, there is some level of social stratification, often articulated with age. The most important criterion for leadership in an Aboriginal community is knowledge. Such knowledge is built up over a lifetime. It is the older people, the elders, who possess the most knowledge, have the greatest assertiveness in articulating knowledge, and have the most intimate knowledge of their own country (Rose 1992: 122). "Old people," Rose writes of the Yarralin, "become increasingly like Dreamings, singularly fixed and enduring. While maintaining their far-reaching ritual obligations, they also become focused on the country in which they wish to die" (1992: 122). The choice of a place of death, then, is a final statement of a person's own identity.

The Power of Sorcery

Another dimension of power/authority in Aboriginal society is that of the medicine men and sorcerers. Even though Aboriginal religious practices have been degraded over the last two centuries, the "Clever Fellows," even in urban areas, continue to be both respected and feared, emotions that can extend beyond the life of the sorcerer. The powers of these men, taken to be supernormal, are derived from two sources: first, from the cult-heroes, spirits, and other entities from the Dreaming; and second, from a long line of medicine men, those who have orally passed their sacred-secret knowledge from one generation to the next. They all have received special training and special initiations and are called men of "high degree." These "Clever Fellows" have much to do with temporality. They are, for example, much concerned with the animistic, the magical, and the super-ordinary causes of illness and death (Elkin [1945] 1977: 37–69).

The men of high degree are believed by Aborigines to do numerous super-normal things, all of which are psychical in nature. These practices include white and black magic—healing and killing by sorcery; use of the "strong eye" for "divining" a murderer or "seeing" whether a sick person's spirit is present; and the practices of telepathy, telesthesia, clairvoyance, and sky visiting. These and other feats are held to require psychic training into the development of a nonordinary state of attention. Illness and death in traditional Aboriginal society are apt to be seen as result of breaking a rule or taboo, a violation of the Law. Through sorcery, health can be restored to the individual, and equilibrium can be restored to the collectivity. Bringing about illness or death can be accomplished by implements such as bones, which can be pointed to the victim while concentrating on negative thoughts about the victim. Once a person grasps that he or she has been pointed, or "sung," he or she is apt to sicken and die unless cured by a second medicine man. The patient, to be helped, must be in a state of mental receptivity and high suggestibility.

Elkin ([1945] 1977: 42) suggests that the use of pointing bones and other forms of projective magic implies a belief that a person can send his or her thoughts through space and into the mind of another person, causing the receiving person to act on, or be at the mercy of, these thoughts. At the extreme, it is believed that a person can be killed by this method. Elkin reports on such a practice in the Lower Murray River region, where a stick fastened to an object that has been in contact with the intended victim (the *ngathungi*) is used for "singing." The person using the object must possess the power of concentration and communicate without physical means. Elkin provides the following example:

> When using this magical implement, the operator takes it in his left hand, and says very softly, "Shooh Ho! Let the breath leave thy body, O boy!" Then he chants a song of hate for an hour or more, after which he warms part of the *ngathungi*, concentrates his mind until he sees a picture of his victim, and then with all the emotion and energy he can summon, he whispers "Die!" ([1945] 1977: 43)

Such a performance might be repeated several times, at least once being physically close to and also "seeing" the body of the intended victim while he or she sleeps and visualizing the intended result.

The practices of "Clever Fellows," even among detribalized urban Aborigines far from their own language areas, are taken seriously and can be a cause of anxiety or hope. The explanation of the cycle of life and death of traditional Aborigines depends both on reincarnation and in a belief in psychic powers. Aborigines see that their life can be threatened by practitioners of these secret arts. It is not unusual for "part Aborigines" with no tribal experience to attribute illness and death to such magical practices, to the activities of spirits, usually of men who have died, and to the breaking of the Law (Elkin 1935).

Authority Ranking in Modern, Western Society

We have seen that in Aboriginal society, tales of the Dreaming stand as a commentary, or a statement, on what is thought to be permanent and as the basis of the world and of life itself. The Aborigines, in Stanner's terms, possess "a poetic key to reality..., a key to the singleness and the plurality of things set up once-for-all when, the Dreaming, the universe became man's universe..." (1979: 29). The Aborigine does not ask himself or herself philosophical-type questions: What is "real"? What are the properties of "reality"? Stanner further notes that this objective questioning of the world "... is the idiom of Western intellectual discourse and the fruit of a certain social history..." (*ibid.*). Time is deeply taken for granted in the everyday world of modern society, being deeply embedded in the social habitus. This fact, however, does not mean that time is of marginal importance to social ranking and social control. On the contrary, the temporal aspect of culture, in Hall's understanding of culture, is a form of communication that "controls behavior in deep and persisting ways, many of which are outside of awareness and therefore beyond [the] conscious control of the individual" ([1959] 1973: 25, 31).

Schedules are the means of temporal control within bureaucracies and formal organizations of all kinds. Schedules contribute to the routinization of short durations through endless repetition (by day or week). They also confine human productive activities by delimiting goals. Schedules involve authority by virtue of the fact that some agents must construct and then implement them, in the process appropriating the time of others. Schedules are used to impose discipline on workers and students and are instruments of surveillance (Rutz 1992: 5). The relationship between authority, scheduling, and the temporal discipline of the body in total institutions has been studied by Goffman (1961) and Foucault (1977).

A Heideggerian ecstatic-futural temporality is a fundamental capability of human consciousness, a form of consciousness that has been developed in an exaggerated form in the modern, Western world. Such conscious action is, in most of its forms, teleological insofar as it plans to achieve and is situated not in the here and now but in the future. On this, Brandon

(1965) writes that the ability to conceive and work toward future ends "is surely the cause of mankind's success in the struggle for existence against other forms of life. The exercise of this ability involves drawing upon the memory of past experience, to guide present action in planning for future situations" (8).

The immediate, first realization of the adaptability of time consciousness is the concept of the future. In order to adapt to, and resolve, the problems that life presents, it is necessary to look to the future with careful planning, to project the likely outcomes of various courses of action, a getting ahead of oneself that enables anticipating, anticipatory resoluteness, commanding and controlling, and editing, monitoring, planning, willing, and intending. This forecasting mind cares about itself, about others, and about the world. To adapt to changing circumstances is to take care of oneself. It is Heidegger's claim that the deepest level of temporality is found in this active, futural orientation and in the processes of working on problems in a productive and an adaptive way. "The primary phenomenon of primordial and authentic temporality," Heidegger wrote, "is the future" ([1927] 1962: 378). Our designations of care and caring—that it can be "ahead of" things, or lay a trap—do not refer to now points in actual clock time but must be understood in their temporal meanings and structure. The future is neither "out there" in space nor hypostatically "in front" of *Dasein*. The future is rather "inherent in the care-structure as its self-approaching process" (Heine 1985: 115). Also, the past is not left behind as a sequence of receding now points.

Authority ranking is always oriented to the future, to the command and control of the future. In Western civilization there has developed a futural time consciousness. Most definitions of education, for example, incorporate, either implicitly or explicitly, the concept of preparing the student—by transmitting information, knowledge, and civic values—for effective participation in his or her future life. Bourdieu (1977: 63), however, suggests, with good reason, that we refrain from any *facile* generalization about modern, Western societies having a future-value orientation based on the notion of progress and a view of the present as preparation for the future (Doob 1960; Vernon 1966; Zern 1967).

There are important class and cultural differences in future-time orientation. What Warden (1968) calls the Western "core culture" emphasizes the deferring of gratification of needs and desires. This emphasis requires the individual to set goals at various future times for which he or she is willing to work. This future-oriented time perspective is less than fully shared with those at the lower end of the socioeconomic continuum, and those of various minority group memberships. Frank observes that "Each culture ... presents its own time perspective and emphasizes the necessity of patterning human conduct in its focus" (1939: 299). Spengler made an even stronger statement: "It is by the meaning that it intuitively attaches to Time that one Culture is differentiated from another" (1926: 130). One important characteristic of Western culture is its orientation toward the future, its striving for improvement and advancement, and its concern for what lies ahead.

The future, according to the common wisdom of the Western world, is where we will spend the rest of our lives. Westerners tend to like new things and show a preoccupation with change and overcoming resistance to change. On the economic level, persistent innovation and restructuring have become necessities in corporate enterprises and in public and governmental bureaucracies. In the economy, constant innovation and progress became key concepts. In order to achieve maximum results (rewards, profits), efficiency has become a social norm for people in most walks of life in the modern world.

The timing, sequencing, and rhythmic organization of daily activities in the modern world go beyond being a mere "linear" orientation, for such a notion of time refers not only to the mundane routines of the everyday world but also to the uses of memory and intention along the past-present-future axis, in an effort to socially construct the future. Progress became a key concept, meaning openness to change and innovation—to new ideas, new ways of doing things, and new things. With the concept of progress comes a new way of conceptualizing time, according to which the past is unrepeatable, the present is transient, and the future is both infinite and exploitable. Western civilization has developed, as a key belief, the notion that change and development are desirable, and that future events can be shaped, planned, and scheduled. Members of modern societies strain toward the future. Ordinarily, the future is seen to be ahead of the individual but is not very far ahead because results must be obtained in the foreseeable future. This usually means months, perhaps one or two years, and at the most, five to ten years. Modern time consciousness is expressed in the ideas of forward progress and nonrepeatable moments. For the individual, there is a need to relate aging and the meaning of life; for the collectivity, there is the relationship of history and its meaning, this being the problem of time (Ferrarotti 1990: 98). Thus with modernity comes the ideology of a future-value orientation, so activities in the present are explicitly oriented to the attainment of future goals and objectives.

With the development of the notion of progress in eighteenth-century Europe, "time ceases to be morally neutral" (Gellner 1964: 3). Excellence was no longer just as likely to be found in the past as the future, and there developed an association of the past with the bad and the backward and the future with the good and the progressive. While this philosophy of breaking with the past has not been shared by all, the Romanticists being a notable exception, it has come to permeate ordinary discourse and thought. Gellner (1964: 3) sees life in the modern world as having come to be lived on an upward slope, in which improvement is anticipated and required. Young (1988), in this connection, invokes the metaphor of time as both a line and an arrow: as a line, time is visualized as a strung-out line, with the two ends representing the future and the past; as an arrow, the metaphor is that of the Arrow of Time shot forward by a "brawny archer" with sufficient force to fly indefinitely, with the arrow representing the future, and the past not a circle but the "... more or less straight road of life's journey" (M. D. Young 1988: 12–13).

Chapter 13

Market Pricing and Ordinary-Linear Time Consciousness

Market-pricing social relations are a human extension and generalization of Plutchik's ([1962] 1991) Territoriality. Plutchik associates with Territoriality a positive primary emotion referred to as "expectation" and "anticipation" and a negative primary emotion, "surprise." Exploration is carried out mainly to exploit the resources of the spaces and places under control. Anticipation includes in its meaning the prediction of supply and demand and the prices of commodities, including stocks and bonds, which can be a key to economic success. This correlation of systems of production, types of exchange, and social relations was well described by Tönnies in 1887 and by Mead in 1934. Mead's concept of "economic relations," which corresponds closely to Fiske's (1991) concept of "market pricing," links people through trade and money.

Market Pricing and Ordinary-Linear Time Consciousness in Modern, Western Society

In this chapter, we consider Western society first and the Aborigines second. In this comparison, we immediately confront a radical difference between culture and nature. The traditional Aboriginal culture sees the person as coming from, being part of, the land and country. Aborigines experience a totalistic involvement and harmonious balance with the forces and life-forms of nature. Nature is humanized, and the human is naturalized. In modern, Western, capitalistic society, culture has rather been conceptualized as the antithesis of nature. All of nature is seen as a resource available for conquest, "development," and economic exploitation, so the land, as in the Australian case, is partitioned into precisely measured lots of real estate, which can be bought and sold as commodities in the market economy. The Aboriginal approach to land is through social and religious institutions, whereas the Western approach to land is individualistic and materialistic (Doolan 1979: 161).

As human social life has developed from feudal and agricultural societies to modern, urban, and industrialized societies deeply involved in trade and commerce, time has become a precious commodity, to be measured, counted, and used. In the business world, the difference between success and failure is heavily dependent upon one's ability to use time to its fullest advantage. Most important is knowledge of the "right time" to buy and sell, based on changes over time of supply and demand for commodities, and of the ever-changing values of stocks, bonds, and currencies. Adam Smith ([1776] 1979) favored the principle of a division of labor in the production process, because it "saves time" and thereby increases profits or reduces losses.

Implicit in this notion of time as a commodity and as the measure of work is that time becomes both linear and quantitative, insofar as labor is extracted from workers by time measured on a ratio-level scale, and workers are paid by the hour, day, or month. This abstract time is homogeneous, objective, measurable, and infinitely divisible (Hassard 1990: 12). Thus time becomes a scarce resource, potentially consumed by a vast number of competing claimants. Time now has value, the scarcity of which has enhanced its worth. Under capitalism, time and money become exchangeable commodities (Lakoff and Johnson 1980). Money can be used to buy time, and money has developed a future value, while time can be invested now to yield money at later times. Once time was created as a commodity, it became an ineradicable reality of industrial social life, which in turn structured the nature of control of the timing and temporality of everyday life. The advent of capitalism required a form of instrumental reason, in Gosden's terms, "to manipulate the world and through which the world was seen as raw material, something to be taken advantage of and used" (1994: 43). Heidegger ([1927] 1962) described a "production mentality" that has developed with modernization. Whereas in the medieval period material things had been viewed as the creatures of God and to be used only with feeling and piety, the things of the world, including human beings who could be economically exploited or even made a *corveé* labor force, became raw materials to be taken advantage of, to be used as mere commodities.

Reality needed to be represented correctly through the operations of the mind before it could be manipulated effectively and profitably. Thus industrial capitalism led to a mechanical and an exploitative view of time, so time itself became not something to be prized or dwelt within but rather to be controlled, regularized, maximized, earned, spent, made up, squandered, valued, wasted, paid for, saved, and fought over. As Lestienne put it, "We not only live in time; we consume time. Time thus conceived of as a commodity confers on it a sense of tangibility, resulting in a reification of clock-time" (1995: vii).

The modern, Western conceptualization of time, articulated with economic processes, stresses speed and temporal precision. All machines have systematic order, and their processes involve sequencing, rate, timing, and duration. Machine time is limited and finite, because it excludes the process of becoming and repetition with variation. "It does not create time in the present but *is* time" (Adam 1995: 52, emphasis in original). In technology, there is an ever-present economic incentive for devices and machines to perform a standardized, repeated task with greater speed, reliability, and energy efficiency.

With the advent of the Industrial Revolution came a need for temporal coordination on a large scale. In the early stages of industrialization, capitalists and managers imposed a standardized clock-time discipline, which was to the workers an alien rhythm that took no account of the natural rhythms of activity and rest of the human mind and body and its relationship to the various rhythms and cycles of nature. Early industrial disputes were of the duration, length, pacing, and sequencing of activities of the working day, and of break time, vacation time, and overtime. The imposition of industrial time, which became a model of all modern complex institutional and bureaucratic forms, was a historically novel expression of industrial capitalism (Thompson 1967; Adam 1990: 110–20).

With the advent of industrial society, time was divided into more precise and standard segments viewed in a straight line, extending back into the past and forward into the future. Factory production required the synchronization of machines and workers. If a group of workers in a factory was late in completing a task, then tasks further down the production "line" would be delayed, and the productivity of workers was thereby reduced. Thus punctuality became a social necessity, and clocks and watches began to proliferate, becoming commonplace in Britain in the 1790s. Hierarchically organized, complex social organizations are powered and controlled by prioritized temporal ordering, synchronization, and regulation as specified in timetables, schedules, and calendars (Adam 1995: 76).

Mumford ([1934] 1955: 5) declared, "The clock, not the steam engine, is the key-machine of the industrial age." An exaggeration, no doubt, but it is nonetheless true that the modern, Western conceptualization of time, articulated with economic processes, stresses speed, precision, and tempo (Nowotny 1975: 329). Clock time, by linking labor power and machines, created the standardized, metronomic rhythms of industrial work. All machines have systematic order: their processes and operations involve sequencing, rate, timing, and duration. There is an ever-present economic incentive for the invention and development of devices and machines to reliably perform a standardized, repeated task. According to Reason, "The history of capitalism can be told in terms of the increasing incorporation of the timing of work operations into the structure of the machinery used in production" (1979: 229). Articulating measured space and measured time, machines parts came to be standardized and precisely made; with time, it became possible to sequence events in terms of microseconds, and with the advent of computers, in terms of nanoseconds and picoseconds.

Labor, in a modern economy, "presupposes a superordinate, abstract, and reified temporal framework" (Reason 1979: 229). Wage labor, explicitly or implicitly, became subjected to a time rate (Marx [1867] 1976: 683–92). Work time was gradually separated from the subjective experience of work, resulting in the dissolution of the qualitative character of working. Time became an abstract medium through which labor was translated into an exact exchange value (Giddens 1981: 130–34). Disconnected from the substance of being and social life, time came to be perceived as real, objective, and associated with money (Nowotny 1985; Thompson 1967). Work time became a substantial, important part of the times of everyday life, being "bargained for, sold, and controlled" (Adam 1995: 26, 85). Linear, clock time reaches its fullest expression in the urbanized, northern countries of seventeenth-century Europe as they developed competitive, capitalistic economies. The rhythms and timing of work involved in commodity production gradually came to be controlled by clocks and schedules. Work in industrial society came to be fundamentally dependent on clock time (Marx 1887; Weber 1904–1905; Thompson 1967; Postone 1993). The commodification of time had become an integral and ineradicable aspect of social life in the modern, industrial world. Thus there is strong evidence of a linkage between market-pricing social relationships and ordinary-linear time consciousness in modern society.

Market Pricing and Ordinary-Linear Time Consciousness in Aboriginal Society

Halpern (1973) compared "whites" and "blacks" in the United States. He describes the disadvantage of blacks as beginning in the institution of slavery, plantation life, restriction to fieldwork, and serflike conditions. Certainly such a comparison also can be applied to the Euro-Australians and the Aborigines in Australia. Far from being prepared to compete in this socioeconomic system, Aborigines in their own culture were not oriented to "getting ahead." Theirs is not a Western notion of time and space. The Aborigines were not prepared for competition with Europeans that began with first contact. Since then, their culture and way of life have been systematically degraded. Even by the 1860s, according to Stanner (1968), about half of the 600 or so tribes had been more or less obliterated. The hard facts are that the rending of their families in the name of assimilation policy—and their loss of lands and resources for ensuring a stable future—was not accidental but rather was, to a great extent, a matter of policy. The result, on the level of personality structure, is well described by Stanner (1968, cited in Stewart Harris 1972: 10):

> ... the apparent mildness and passiveness of the Aboriginal today is a product mainly of four things: hopelessness, powerlessness, poverty, and confusion; that there is no more terrible part of Australia's nineteenth-century story than the herding together of broken tribes into artificial settlements and institutions, under alien authority.

What must be inferred from this statement, and from the persuasive literature on which it is based, is that the Aborigines' damaged social institutions, health condition, poverty, erosion of religion, and so forth have been a shattered world. While Aboriginal society and its distinctive ontology were being devastated, in modern, Western, capitalistic society a linear time that is alien to Aboriginal culture has gradually become hegemonic.

Kolig (1981) reports that in the Fitzroy River area of the Southern Kimberleys there is considerable trade or barter in complex, mythico-religious traditions from one group to another, these exchanges being facilitated by the Aboriginal "gift" or exchange system. In this "religious business," religious objects and ceremonies are seen as merchandise. In this trade, Aboriginal religious items are rapidly handed along on well-established routes as communities, moving 100 or more kilometers apart, being exchanged in a fast and lively process. Before contact with the

Europeans, no Aborigine was landless, because everyone inherited a stretch of their ancestors' songs and the route that their most significant ancestor walked and over which the songs were sung.

Barter and exchange historically took place between tribes throughout Australia, involving items such as boomerangs and acacia wood for making boomerangs, spun hair and human hair belts, shell ornaments, ocher, bamboo spears, and a narcotic, *pituri* (*Duboisa Hopwoodi*). The location of the trees from which *pituri* is made and the secret of its preparation were secrets jealously guarded by tribal elders. Doolan (1979) writes that "Other tribes did not travel to the area where the *pituri* was obtained, but the guardians of this item took it from the country many miles distant to a central 'market place', where the other tribes met to exchange various implements not readily available to the '*pituri* people'" (164).

The division of labor according to age is key to understanding economic production in traditional Aboriginal society. There are those too young to participate in production, those who do produce, and those too old to contribute to production. In a study of tribal-living Aborigines, Rose (1987) found that males were at their peak productiveness at about age forty, when they were most likely to be married and consequently having the greatest responsibility for providing protein food for their families. The younger, unmarried men were the most skilled hunters, whereas the older men were the best organizers of production and distribution. Rose (1987: xv) draws on Meehan's (1982) study of shellfish gathering to infer that Aboriginal women reached their peak productivity at about age forty-five, the inference being that younger women were inhibited in this activity by the burden of looking after children. Aborigines produced little beyond their immediate day-to-day needs, so there was no significantly realizable surplus product.

December 7, 1941, the day the Japanese attacked Pearl Harbor, was a watershed for the Aborigines. From that day forward, tribal-living Aborigines were progressively involved in the cash economy of Australia. Participation in hunting and gathering was reduced. Rose (1987: viii) suggests that it would be a delusion to equate the stable, almost sedentary groups of Aborigines living on outstations with traditional hunting bands or foraging groups. Yet Aborigines, even in urban areas, continue their hunting and gathering of foodstuffs and herbal medicines in a way that is indeed impressive.

Once Aborigines came into contact with, and under the control of, Europeans, they, out of necessity, tended to adapt a sedentary way of life, eking out an existence on the fringes of European towns and cities and

FIGURE 13.1 A famous image of Aboriginal life around Sydney in the early 1830s. This watercolor by August Earle captures the profound effect of dispossession. The English settler and his wife look out from their verandah, while the Aborigines huddle around a fire on the settler's lawn.

finding work on cattle stations. In the southern and eastern parts of Australia, Aboriginal people had been almost fully dispossessed of their lands before the end of the nineteenth-century (Harris 1990: 549), as illustrated in figure 13.1. Christian missions became places of physical survival but of cultural dispossession. All of these factors brought about a profound erosion of traditional Aboriginal culture. On the change in Aboriginal family organization, Elkin writes: "The custom of old men marrying young girls is changing in those parts of northern Australia occupied by whites. Many young men are now seen with young wives though seldom with any children. Polygamy is also being dropped" (1954: 127n). Before World War II, Aborigines in the north came into contact with whites, primarily with cattlemen, missionaries, and policemen. The Aborigines, demoralized and socially disorganized and divided into small and isolated groups, were susceptible to the disapproval of what O'Malley (1989) and others referred to as the Aborigines' "morally evil" traditional customs and laws. But after this war, Aborigines were no longer viewed as a "dying race"; rather, they were developing a Pan-Aboriginal political consciousness. Their immediate dependence on Euro-Australians was lessened, and they were able, to a limited extent, to partially reintroduce their customs and cultural forms in the name of fostering Aboriginality. Contemporary Aborigines differ greatly in the extent to which they are "tradition oriented" and can choose to live in the bush, near outstations, in fringe settlements, on reserves, or in towns, suburbs, and cities. Today even the country-oriented Aborigines, F. G. G. Rose points out, "are not living under traditional economic conditions. They are inextricably involved in a money-commodity economy" (1987: 16).

In terms of economic production, the single alternative to hunting and gathering that has been "available" to the Aborigines over the last two centuries since first contact has been the Euro-Australian version of the modern, industrialized, capitalistic society with its cash economy. In their relationship with the dominant, modern society, Aborigines lost free access to their lands and to their economic base and became, to an extent, incorporated, initially through coercion, into the cash economy as proletarian laborers and service workers. This mode of economic production has meant many things foreign to traditional Aboriginal society: the introduction of large-scale, hierarchically organized agencies and bureaucracies and participating in an economy and a society spatiotemporally organized by two quantitatively interdependent ratio-levels scales, those of money and time.

Radcliffe-Brown, in his study of the Pintupi and other Aboriginal groups, argued that their social organization consisted of two fixed groups, the family and the "horde," the horde being "a small group owning and occupying a definite territory or hunting ground" (1930: 34). The claim, by Radcliffe-Brown (1930) and Sansom (1980), that Aboriginal society is organized around corporate groups formed on the basis of valued property or real estate is, as Myers (1986) clearly demonstrates, not descriptive of *tribal-living* Aborigines. Myers shows that for the Pintupi, "things"—prerogatives, duties, ritual, objects, land—all have cultural meaning, largely as expressions of both autonomy and relatedness (shared identity).

Woodburn (1980) locates traditional Aboriginal concerns for enduring relationships in the bestowal or the promise of women, but for Myers the basis of temporal continuity is to be found in the process of objectification. Propertied objects—tools, clothing, food, automobiles, and so on—are easily obtained by the Pintupi, these objects then moving sometimes with ease and sometimes with difficulty through networks of overlapping and nested relatives, friends, clan members, and tribal members. The importance of such objects is that they provide the opportunity to say something about oneself, to give to others, and to share. Rights to objects enter into a system of exchange that constantly negotiates relationships of shared identity. We will consider just two of Myers's several examples, land ownership and the ownership of motor vehicles.

Landownership

Among the Pintupi, landownership is not a kind of personal "property." It is rather the case that shared identity is readily extended to others in the form of recognizing an identity with named places and sites. Landownership, Myers (1988: 65) explains, is a negotiated outcome of individual claims and assertions. There is an emphasis on sociability and negotiation, which underlies individual claims to be affiliated with multiple landowning groups. Myers concludes, "Landownership is not primarily an ecological institution but rather a political arena in which the Pintupi organize relations of autonomy and shared identity" (1988: 65). Ownership consists primarily in control over the stories, objects, and rituals associated with their mythological ancestors. Access to this knowledge of these esoterica and to the creative essence they contain is restricted, and one can acquire it only through instruction from those who have previously acquired the necessary knowledge (1988: 65). Landownership for tribal-living Aborigines is of a contractual nature, the basic kind of contract being that of marriage, which extends time both into the future and the past. Among the Yarrilin, marriages are contractual arrangements made between groups. They are intended to repay or create new obligations, to initiate future obligations, and, more generally, to prepare a new generation that will assume responsibility for its country when their parents die. Marriages are arranged with country affiliation and responsibilities in mind (Rose 1992: 126).

In a landowning group of Aborigines, members are responsible for "looking after" the stretch of country that they hold as a hallowed trust. Their country comes to them from a sacred past, and they are, in turn, expected to pass it on, intact, to their own descendants. A primary obligation of every generation is to produce a new generation able to take responsibility for its country. Looking after the land is the responsibility of fully initiated adult males—which means caring for the land in the physical sense but also, as described earlier, by the performance of rites, prescribed from the past and related to the mythical characters that were linked to that part of that country (Berndt and Berndt 1978: 20). Ownership of a country is never absolute. The owners of land are in a position to make possible the exercise of equality with other fully adult persons, to offer ceremonial roles to others (as part of an exchange), and to share rights in ritual paraphernalia. At the same time, people view "country" (*ngurra*) as the embodiment of kin networks and as a record of social ties that can be carried forward in time.

There are at least two instances of post-contact struggle over land between Aboriginal tribes. D. H. Turner (1985: 84–85) describes an intense Pitjantjatjara-Jankudjara struggle over resources and an annexing of territory, a situation that was in part caused by drought and by the destructive effects of introduced animal species. Meggitt (1962: 42) describes the Warpiri-Waringari war that took place at Tanimi, which was a rabbit-infested, gold-rich area suffering from drought and postcolonial pressures. Each side sustained about a dozen fatalities in this contest. These two exceptions prove the rule. The Aborigines have historically been peaceful people, a fact that stands in sharp contrast to the long history of warfare that has plagued Western civilization.

Motor Vehicles

Among the Pintupi, motor vehicles are used for transporting people, getting supplies, for visiting friends and relatives, and for hunting expeditions. Ownership of such a vehicle can fall under two categories of "property," in English, "private" and "community"/"company." The vehicles provide people with autonomy, such as the ability to embark on a hunting expedition whenever there is a wish to do so. But this autonomy is limited, because possession of a vehicle makes one the target of requests that set in motion a multitude of social ties and obligations. As Myers notes, "to have a car, one might say, is to find out how many relatives one has" (1988: 55). Those without a car constantly pressure others for help with transportation. Many disputes are occasioned by relationships to motor vehicles. Owners of vehicles are sometimes relieved to have their vehicles break down, only then being free of problematic, contradictory, and sometimes impossible-to-meet demands. Such disputes among kin have on occasion led "proprietors" to set fire to their own vehicles. But car ownership has a positive aspect as well, providing an opportunity for a person to give to and help others, which can result in respect and appreciation for having "looked after" kin and other community members. Aborigines regard such personal possessions, even expensive cars, as replaceable, much as they do spears or digging sticks. Thus they say even when a $4,000 car is destroyed after only a few weeks of use, "There are plenty more motorcars; no worries" (Myers 1988: 61).

The Pintupi, and other Aborigines as well, clearly are not "materialistic"—they much prefer to invest their time and energy in people rather than in things. "The accumulation of private objects," Myers adds, "... is not the means through which one's identity is transmitted through time. This is secured, it would seem, in part, by giving such things away" (1988: 62). Because it is difficult to refuse help of various kinds to others, "private" cars are the basis on which shared identity is constructed. Myers reports that the Pintupi come to regard a particular vehicle as representing a cluster of associates that, often for the life of the car, travels together. Myers provides an example: "A certain blue Holden is 'that Yinyilingki motor-car', identified with a group of young men. The car is the occasion for their relationships and obligations with each other being temporarily realized.... In short, it becomes an objectification of a set of social relations" (1988: 62).

A second category of motor vehicles is the "community" car, which can be a source of confusion and conflict. Such vehicles usually come from governmental or foundation grants. The problem of property is assimilated to an ambiguous notion, that of *kanyininpa* ("to have," "to look after"). Thus, for example, a particular councilor might assume the major responsibility of "looking after" a certain vehicle. But this Aborigine is not the vehicle's "owner." Should he refuse to give rides, then others might complain that he is "jealous" for the vehicle, is "selfish," and is not "looking after the people properly." The vehicle will be described as being "for everyone" and is intended to "help people" (Myers 1988: 63).

Religious Knowledge and Sacred Places

Knowledge of sacred places—where the entities of the Dreaming carried out great deeds, and which are the locations for important ceremonies in which the stories of these deeds are retold—is highly valued and is vital to cultural reproduction. The religious interests that an Aborigine has in the lands to which he or she belongs are socially sanctioned, expressed in performances commemorating the exploits of creatures, such as leading songs about these lands and ancestors of the Dreaming, and also in the transmission of that knowledge to others. Religious knowledge is of utmost importance in tribal society. The religious tenets and ritual actions are taken to be the embodiment of spirituality ordained during the epoch of the Dreaming. These tenets and actions are inviolable and sacred. Religious ideas are articulated in stories, songs, dances, rituals, bark painting, wood carving, and a myriad of other artistic forms.

In times of extreme poverty, the Warmun people of the East Kimberleys, according to Ahern, "used cardboard boxes, pieces of tin, masonite, even fly screens found in the local dump, as the ground on which to paint" (1991: 34). These works of art, displayed during corroborees, have spiritual matters as their topic. Aborigines show great determination to involve themselves in their cosmology and to artfully express their worldview. The various forms of art are often combined in major religious events, these events often being referred to in English as "business." Sacred-secret religious knowledge is unevenly distributed but is valued by all societal members, even where Aboriginal culture has been seriously eroded, for example, in the western suburbs of Sydney. Cultural knowledge is rather like a commodity that can be used in exchange transactions that have clearly defined social consequences. The specific forms of such knowledge—songs, rituals, and so on—can be thought of as "commodity classes." Restricted access to such knowledge inhibits "supply," which makes such knowledge "valuable." The rationale for restricting religious truths in Aboriginal society is that these truths are so spiritually powerful that they could cause real harm to those unprepared for exposure to this knowledge and the responsibilities that it brings. For example, the sponsors of adolescent boys undergoing the circumcision-initiation process prepare them in a highly structured manner. The religious knowledge a tribal Aborigine receives is revealed successively over time and is a lifelong process.

It is the possession of this knowledge of customs, rites, and, more generally, The Law that explains the high status of the tribal elders. Gradually, when the Aborigine has progressed through a long series of knowledge-conferring rites, ceremonies, and practices, he will assume the role of instructing his younger apprentices and charges. Males who are not initiated and who do not participate in ritual life are, by many tribal Aborigines, considered ignorant and apt to be called "boys." Thus differential mastery of religious knowledge contributes to social differentiation. Continuing the market-pricing analogy, those who receive religious knowledge owe a "debt" to their instructors. These debts can be "paid" in various ways. For example, in initiation rituals, the parents of the youth can promise a daughter to be the circumciser's wife. In other rituals, the initiated person might present to his instructors a hunted kangaroo. The "paying" for secret-sacred knowledge through such exchanges lasts for the duration of the relationship between inductee and inductor. There are, of course, many other kinds of indebtedness and forms of exchange, with considerable intertribal variation (F. G. G. Rose 1987).

"Holding a country" (*kanyininpa ngurra*), the decidedly non-Western, Pintupi notion of "ownership" of their tribal lands, provides opportunities for organizing significant events, ceremonies, and rituals. The Aborigines responsible for these events are not called "owners" but rather "custodians" of the land and its sacred sites. These custodians, usually initiated elders but not necessarily males, are able to grant ceremonial roles to others and also to share the rights to use ritual paraphernalia. The mythologization of such places is never really finished, for it requires repeated interaction rituals to be carried out at the appropriate times of the year, with the result that the Pintupi, and Aborigines in general, come to view their "country" (*ngurra*) as the embodiment of social networks of kinfolk and tribal members and, Myers adds, also serves "as a record of social ties that can and must be carried forward in time" (1988: 65).

Swain (1993: 51) similarly argues that the ontological identity between a human being and an Abiding place, a sacred site, does not provide the person with exclusive social, political, and economic rights to the place. Writing of the Walbiri, Swain explains that no person has exclusive rights to his or her site of origin, and that most persons have limited rights to a multiplicity of sites. A sacred site can, in fact, be more dangerous to the person who springs from it than it is to others who are in a different and less important relationship to the place. The danger emanating from one's own place necessitates social interdependence, alliances, and relationships. On the economic level, there is no straightforward relationship between identification with a site and "ownership" of that site. Economic autonomy is always counterbalanced by relationships to other people, whose relations to a variety of sites must be respected. Responsibility of "holding the country" weaves places, the paths that connect them and their people into what Swain calls "an interdependent network" (1993: 53). The mutual interdependence of places and their "clans" makes possible not only tribal cooperation but also intertribal social cooperation. Under these conditions, there is no economic exploitation of one clan by another, or of one tribe by another (T. G. H. Strehlow 1956).

Personal Property of the Deceased

In death, the Pintupi do not regard personal effects as an estate nor as personal mementos to their heirs closely identified with the deceased. For the Pintupi, things associated with the dead person make them sad, so that such things are effaced, given away to distant relatives, preferably of the mother's brother kin category. Objects not given away are apt to be destroyed. Motor vehicles, for example, might still be seen occasionally if given away to such relatives. Rather than risk a continuation of grief triggered by the sight of such a vehicle, a likely course of action would be to burn the vehicle. The camp or house that the deceased inhabited is likely to be sold or abandoned, partly out of apprehension of the person's spirit or ghost continuing to inhabit the camp or dwelling. The place of burial might be avoided for several years. The personal name(s) of the dead person and anything sounding like it are generally avoided and substituted for in everyday speech. The goods and personal effects of the deceased, Myers writes, "are not allowed to carry the deceased's identity forward in time" (1988: 72). Every Pintupi person comes into the world with an association with the Dreaming at a particular place. This provides an assured identity no matter what happens to one's personal possessions, which contrasts sharply with Western cultural practice, in which accumulated personal property and wealth constitute an important medium through which personal identity can be realized.

The resilience of Aboriginal cultural practices has been supported by a stubborn Aboriginal substantiation of the person which, as Berdyaev writes, "integrates the past in the future and eternity" (1938: 100). Sansom conceptualizes Aboriginal personality as being the opposite of Rowley's (1986: 134–35) agentic individual as he writes:

> Aborigines are persons who are particularized into existence and are, in themselves, the past emergent in the present. Where identity is thus always emergent, there is no succession but the recognition of the particularity of death's undoing which is symbolized in the prohibition that forbids the calling of the names of dead people. (1988: 158–59)

There is thus in Aboriginal society no assurance of personal continuity. Continuities are rather emergent phenomena born of a way of conceptualizing the present in relation to past histories of consociate experience. A consequence is that when people die and their spirits have been ritually returned to country, there is a collective experience of final loss.

Negative and Positive Experiences of Market-Pricing Social Relations

Tribal-living Aborigines must cope with the Australian cash economy and the economic marketplace, and for many Aborigines, market pricing includes working at a full-time job, making mortgage payments on a house, making payments on a motor vehicle, paying for insurance and registration, and, in addition, possibly obtaining various kinds of governmental aid. Aborigines are generous in their contributions of time and energy to Aboriginal organizations—such as health centers, homes for orphaned children, shelters for women, and early family educational programs. Partially assimilated suburban-to-urban Aborigines now wear watches, catch trains on a clock-time schedule, and increasingly have nearly the same distribution of work time and leisure time as do non-Aborigines living in modern Australia with its cash economy. To the extent to which Aborigines are involved in the sociotemporal order of modern society, they will develop and make use of a linear time consciousness.

Mead (1934) and Durkheim ([1893] 1933: 392–93) hold a benign view of market pricing, focusing on the positive experience of this social relationship. However, there also exists the negative experience of market pricing, that of being economically exploited. Aborigines have historically been subjected to extreme forms of economic exploitation. A central aspect of their negative experience of market pricing is the loss of their traditional lands and territory. An equally central aspect is the exploitation of their labor, summed up by the statement of one life-history informant: "If we weren't up for work by 5 A.M., we'd get a hiding." A second example comes from an Aboriginal woman's life-historical testimony about her work experience after being "taken away," as a child, from her family and home:

We got paid	and they'd give us the two shillings
for working there,	and make us slip it in.
two shillings a week.	See, we'd have ten minutes
But if we did anything wrong	to wash up the dishes,
They'd take that two shillings	and if we didn't do it in time
off of us	we'd lose the two shillings.
and put it in	All of our chores were like that.
the Lady Lawley Cottage Fund.	If we didn't do it by a certain time
They had this tin with	our money would go into the tin.
Lady Lawley Cottage written on it,	

In chapter 15, data pertaining to this question will be presented, indicating that the positive experience of market pricing is predictive of ordinary-linear time in both cultural groups, and the negative experience of market pricing—as illustrated by the aforementioned testimony of the imposition of linear time—is predictive of linear time for Aborigines but not for Euro-Australians. It also should be noted that many aspects of market-pricing social relations are neither positive nor negative but neutral.

Six Propositions: A Look Ahead

We have now linked each of the four elementary social relations to one of four elementary forms of time consciousness. It has been argued that these four social relations are *culturally universal*, existing in every human culture. We have seen that while traditional Aboriginal culture stresses communal-sharing and equality-matching social relationships, it is not devoid of authority ranking and market pricing that exist in a minor form but exist nonetheless. It also has been argued that the four kinds of time consciousness are *cognitive universals*, aspects of the four most global forms of information processing made possible by the overall organization of the human brain.

We also have seen that there is a close connection between communal sharing and equality matching, so they find a veritable fusion in hedonic community. A similar fusion of authority ranking and market pricing social relations results in an agonic society. But there is no evidence whatsoever that the hedonic and agonic forms of community and society are culturally universal. They are not. These two concepts were developed in comparative primate ethology, in which it has been found that hedonic society characterizes nonhuman apes but not monkeys, and that agonic society characterizes monkeys but not apes. In human society, we have examined one culture—the tribal-living Australian Aborigines—that can be said to be fundamentally hedonic in its social organization, and another culture—the modern, industrial, Western way of life, with its cash economy—that most certainly contains hedonic social relationships but is, in addition, agonic.

An argument has been made that there can be a veritable fusion of patterned-cyclical and immediate-participatory time consciousness that results in an even higher-level cognitive structure, which is termed *natural* time experience. And there can be a fusion of episodic-futural and ordinary-linear time, as would appear to be necessary for modern economic enterprise, resulting in a second, highest-level cognitive structure, that of *rational* time experience. Natural time experience and rational time experience are *potentialities* of the human mind, but no claim is made that they are universal.

For the market-pricing/communal-sharing polarity, we have the following two propositions:

1. To the extent that members of a culture participate in social relations that involve territory, property, symbolic capital, and other resources that are valued, priced, or rendered marketable, as agentic individuals they will give emphasis to logical-analytic thinking, including a time consciousness that is ordinary, linear, and based on clocks, calendars, and schedules.
2. To the extent that members of a culture participate in social relations that involve interrelatedness, communality, and collective representations, as a basis of social solidarity, they will give emphasis to gestalt-synthetic thinking, including a patterned-cyclical time consciousness.

And, for the authority-ranking/identity-matching polarity, we have the following two propositions:

3. To the extent that members of a culture participate in social relations that involve power, prestige, influence, rank, authority, and other forms of social hierarchy that are prized, sought after, earned, or otherwise allocated to high-status individuals, they will emphasize episodic, conative information processing, including an experience of temporality that is emotion-laden, episodic-futural, and temporally stretched.
4. To the extent that members of a culture participate in social relations that result in agreement, like-mindedness, consensus, reciprocity, and other forms of social equality that are shared, matched, and agreed upon, they will emphasize the participatory mode of information processing, including an experience of time that is immediate-participatory, present-oriented, and temporally compressed.

For the agonic/hedonic polarity, we have the following two propositions:

5. To the extent that members of a culture participate in positive, agonic society (the unity of the positive experiences of market pricing and authority ranking), their time consciousness should be simultaneously ordinary-linear and episodic-futural (i.e., they should have a rational experience of time).
6. To the extent that cultural members participate in positive, hedonic society (the unity of the positive experiences of communal sharing and equality matching), their time consciousness should be simultaneously immediate-participatory and patterned-cyclical (i.e., they should have a natural experience of time).

We will see, in chapters 14 and 15, that it is possible to measure the four elementary forms of temporality, and that once this is accomplished, it becomes possible to operationally define, and empirically measure, natural and rational time experience. Also, once we have measures of the positive experiences of communal-sharing and equality-matching social relationships, it is a straightforward matter to measure the positive experience of hedonic community by an interaction term; moreover, once we measure the positive experience of authority-ranking and market-pricing social relationships, it is not difficult to measure the positive experience of agonic society with an interaction term. This methodology, then, makes it possible to test the fifth and sixth propositions of the theory.

Our next task is to find an appropriate *kind* of data to examine, find a way to measure what must be measured from the data and then testing the six hypotheses of the theory corresponding to its six propositions. In the next chapter, it will be argued that *textual* data are what we need, and that the best textual data we might seek are to be found in life-historical interviews. A methodology for the measurement of time consciousness and social relations from such text will be introduced in the next chapter and then in chapter 15 further elaborated on and put into practice.

Chapter 14

Text and Temporality

One's memories of life are intimately connected to language. Ever since the invention of the tape recorder, stories told about one's life are verbal accounts that can be transcribed as verbatim texts. The primary origins of such texts are social relations and social events. Our understanding of *biographical time*—with its orientation to both the past and the future—depends upon the use of language. Such understanding requires a temporal stretch that extends the context in relation to a project and a concerned understanding necessary in order for meaning to be communicated and shared. Language provides for liberation from the immediate present and a measure of control over the future. Through language, we can choose from among various course of action. Dewey wrote that language, together with typification, makes it possible for "every experience [to live] on in future experience" and for the past to "provide the only means ... for understanding the present" ([1934] 1958: 344–45). The main value of an account is its power to tell us something worth knowing about our temporal world, by its ability to constitute a narrative (Ricoeur 1983, 1985, 1988). When such an attempt succeeds, it creates what Ricoeur calls "temporality," by which he means the structure of existence that reaches language as narrativity. And as Richardson writes, "We reach ahead toward our ends, from out of our rootedness in what we have been, and through (or by means of) the entities with which we are preoccupied" (1986: 94). An authentic experience of time, then, must appreciate both the past and the future.

The study of individual lives, as bearers of culture, in a general sense, requires the use of text produced by individuals reflecting upon, and telling stories about, their lives and times. From Heidegger's notion of temporality stretched from birth to death as a basis for life experience, it becomes apparent that the texts appropriate for the analysis of temporality and time consciousness require reflection upon one's entire life. The logic of this argument takes us immediately to life-historical interviews and autobiographies as *the* most appropriate sources of data for sociohistorical analysis. The life story represents an overall construction of the informant's past and anticipated future life, in which relevant experiences are linked in temporally and thematically consistent patterns. The consideration of one's life does not mean simply a series of isolated experiences, laid down in chronological order; it "must be rather interpreted in the gestalt sense of biography as a comprehensive, general pattern of orientation that is selective in separating the relevant from the irrelevant" (Kohli 1986: 93). In reconstructing his or her life history, the informant connects and relates events, actions, and experiences with other events, actions, and experiences according to substantive and temporal patterns. These patterns do not follow the linear sequence of objective time but rather conform to a perspectivist time model of subjective time (Flaherty 1991).

Life histories of individuals can be seen as cultural histories in miniature. As Ferrarotti points out, a life history, being manufactured by a real actor within the history of a culture, justifies reflecting on the life course by "biographical testimony and autobiographical accounts" (1990: 107). In the life-historical interview, the informant draws from his or her memory, which makes it possible "to integrate experience in a series of ongoing syntheses which become understandable as we interpret the past and a future in a changing present" (Kern 1983: 45). Thus while a life-history interview is a personal account and a private statement, it also is a bearer of collective meaning.

In an oral history, the experience of the present provoked by the interview itself brings the external temporal structure to bear on the creative moments of past and future penetration, which is crystallized in the creative narrative of the interview and which informs the primordial temporality of finitude (Heidegger [1927] 1962). As explained by Schutz and Luckmann, "knowledge of finitude stands out against the experience of the world's continuance. This knowledge is the fundamental moment of all projects with the frameworks of a life-plan, as it is itself determined by the time of the life-world" (1973: 47). Schutz and Luckmann define their concept of multiple life-worlds in terms of temporality, as *moments* in which the private and the public collective intersect and interpenetrate.

The experience of producing an oral self-history involves confrontation of tension between private and public life, which will likely render the experience cathartic and gratifying for the informant (Kaplan 1982). It would be too much to assert that playing the role of informant in a life-historical interview has a therapeutic foundation, that it is constitute of narrative therapy, but it can certainly have beneficial effects. The telling of one's life history does not necessarily lead to catharsis or acceptance of reality; but if the past is scrutinized with care and understanding, then the result can be a life that is enhanced in its health and authenticity (Kern 1983: 61). As Kaplan writes, "The life history method creates 'extraordinary' people out of 'ordinary' people. In giving people a chance to be really heard—the *significance* of their own histories is revealed" (1982: 48, emphasis in original). The present of the interview, in this context, is not merely the now, the point on a time line but is rather an accomplishment and an act of creation, being as it is the result of complex and difficult action. By gathering and making coherent the past, the informant is potentially strengthened and possesses enhanced resources for formulating and carrying out plans in the future. This strengthening of the individual is reinforced and taken to a new level insofar as members of a community are able to productively share the narratives constitutive of collective history.

The informant's past is brought forward to the present by the socially constructed stock of cultural knowledge, by language, and by the relevance of the person's extended social relationships, all of which are past dependent. According to Adam, this means "first, that the past is fundamentally embodied in projects, and second, that the public and the private, the objective and the subjective, the past, present, and future, interpenetrate in actions and their representations" (1995: 78). Certainly the telling of stories from one's life and time is in itself a significant social interaction and communicates not only the story content but also the *cognitive orientation* of the author. The interview, then, is fundamentally temporal: informants refer both to their common past and their collective future. Adam points out that such narratives "allow us to think in both the past and the future tense: we can conceive of past pasts, future presents, present futures, and future futures, or a number of their combinations" (1995: 78). A life-historical interview necessarily takes place in the present, with the tape-recorded audio record, and subsequent transcripts, existing as valuable cultural artifacts. The reality that emerges in such a present, according to Mead's ([1932] 1980: 1–32) philosophy of time consciousness, reflects into the past, recreating, changing, and developing its meaning in light of the new present, occasioned by the interview as conversation. The past, Mead correctly proposed, is not the past "in itself" or "out there" and has no independent status. Only its relations to the present provide the grounds of its pastness. Mead recognized that the reconstruction of the past was crucial to his interest in resolving the contradiction between determinism and emergence, because "the novelty of every future demands a novel past" ([1932] 1980: 31; also see Flaherty 1991: 76). The past, in Mead's radically insightful and pragmatist view, is as revocable, and as hypothetical, as is the future. The "real" past, just like the "real" future, is unobtainable. Only through the action of mind do we transcend the present and extend our environment. Only through mind is the past open to us in the present. Mead saw that "intelligent control of immediately foreseeable conditions requires grasp of the emergent and novel" (1934: 314, 339). Mead viewed the past as a personal history, arising through memory and existing in images that form the "backward limit of the present" (1929: 235). Similarly, the future is a hypothetical existence. Mead argued that the past is as open to social construction as is the future: "We speak of the past as final and irrevocable. There is nothing that is less so" ([1932] 1980: 95, 66).

Life-historical interviews are arguably the key source of data for learning about, measuring, and testing hypotheses about time consciousness and temporality—through the investigation of biographical text. The interview, ordinarily a dyadic social interaction, is an expression of sociality. Sociality is essentially temporal, and temporality is irreducibly social, because both are located in the interactional process itself. Thus temporality is compressed in the moment of the interview; this moment is not a mere recounting of the past but is rather an emergent phenomenon (Mead [1932] 1980). Time is creative, constitutive, and transformative. In the life-historical interview, there is an interruption of the flowing unity of the informant's stream of consciousness in the search for stories that represent and reflect personal history and the experiences of significant-other persons. The product is not a single, unitary life story but is rather a number of stories told about the informant's life and times. The meaning of these stories is limited by the informant's selection processes, with the unsaid aspects of any story beyond the possibility of analysis. This limitation follows from the nature of temporality according to which the things, processes, and objects of mental attention in the world are never full present nor fully absent, never fully remembered nor fully forgotten.

Primordial temporality constitutes the opening of historical horizons through which the advent of the "world"—the collective involvement of people within society and the emergence of entities in nature—can unfold. Strenuous effort is apt to be involved in an informant's telling of stories about his or her life and times that can, on occasion, leave the informant feeling drained and exhausted. Every act, including storytelling, "involves a span of time and pre process of reflection and self-identification, and thus is not a mere arrangement of isolated moments" (Maines, Sugrue, and Katovich 1983: 161). The process of struggling to remember important past events that may have been repressed or otherwise not thought about renders current action constitutive of a real and an important present. This present requires real exertion and mindfulness as memory is searched and narration is linked to significant acts, decisions, and situations that a person has faced in the past and is likely to confront in the future. Memory is put into the present by the act of storytelling, which allows the present, the past, and the future to contribute to a life history. Such involvement in the interview "makes memory more consistent ... [and] restores it to the practical domain of action" (Minkowsi [1933] 1970: 33).

NEUROCOGNITIVE, HIERARCHICAL CATEGORIZATION ANALYSIS

The methodology to be used for the present analysis is the lexical-level content analysis of text, of all of the words produced by the informant in storytelling in a life-historical interview. To this end, Roget's ([1852] 1977) *International Thesaurus* was used. Roget provided a remarkable hierarchical classification of the English language. He worked consciously in the tradition of seventeenth-century rationalist philosophy, making a nearly heroic effort to map the totality of concepts of the human mind. He worked on this task for forty-seven years. His broadest classification contains eight *classes* of words; abstract relations, space, physics, matter, sensation, intellect, volition, and affection. Under these classes, we find *notions* (e.g., under abstract relations: existence, relation, time, quantity, change, event, causation, power, and motion). Under the notion "time," we find five *categories*: absolute time, relative time, time with reference to age, time with reference to season, and recurrent time. Under these five categories, we find varying numbers of *folk concepts* (key words). Under the folk concepts, we find a list of individual words and phrases.

The method used was to select folk concepts as manifest *indicators* of the four time consciousness variables and the eight variables measuring the positive and negative experiences of EM, CS, AR, and MP. In making a word list from the folk concepts, certain subcategories with meanings tangential to the overall concept were deleted at the outset, and then all possible forms of every word under the key word were considered for inclusion. The primary denotation of every word was used as the criterion for classification and for deciding where to place words that were assigned to two or more folk concepts by Roget. The method used for word list construction was thus objective and could be roughly replicated.

The resulting method is called the "Neurocognitive, Hierarchical Categorization Analysis" (NHCA). We have just seen how it is hierarchical. It is neurocognitive because concepts from brain theory are used to construct indicators for the modes of information processing that are associated with regions of the cortex. For example, the measurement of episodic-futural time consciousness was based on descriptions in the neuroscientific literature, with the following Roget folk concepts selected as indicators: Will, Resolution, Intent, Plan, Foresight, Presentiment, and The Future.

NHCA necessarily assumes a concordance of "experience-to-word-sense mapping" (Alverson 1994: ix) across languages, which is consistent with Alverson's finding that while languages vary in the ways they map words (and phrases) onto experience, meanings can be translated quite exactly from one language to another. Roget proceeded on a similar assumption, as he believed his "large categories of ideas," or folk concepts, are universal across human languages and are capable of being understood by human minds everywhere. From this perspective, then, we can expect that the extent that a person's mentality emphasizes the logical-analytic mode of information processing of the left hemisphere of the brain, he or she should make extensive use of words pertaining to cause and effect, number, sequence, quantity, and ordinary-linear time. And to the extent that a person has a linear conceptualization of time, for instance, we might expect the use of words pertaining to clocks, calendars, schedules, timing, durability and lateness, age in years, and so on. While only words are used in the present analysis, Alverson has demonstrated that meaning is conveyed not only by words in a grammatical context but by collocations—that is, phrases with stereotypical meanings.

Roget developed an inventory of 1,042 "broad classes of words," here termed *folk concepts*, serving as the single source for multiple indicators of the eight social relations variables and the four basic kinds of time consciousness. Under the folk concepts, Roget listed words (and phrases) that were used to generate lists of individual words. In many cases, the various forms of a word were assigned to different folk categories. The division of words into their folk concepts was based on the first meanings of the words and was a *partition*, so words that fit two (or more) concepts equally well were excluded from the analysis, and no word was used in more than one word list.

On the level of the categories of reason, that should be an enhancement in the extent to which there is a use of words, and the ideas they express, for words that are in categories that would seem to indicate the salience of these categories. Consider two examples: we have defined patterned cyclicity as dualistic, so we can expect words that would fit in Roget's folk category "Duality"; and patterned-cyclical time has a "fusion" or an "interpenetration" of past and present that should lead to the use of words that simultaneously fit into two Roget folk categories, "The Past" and "The Present." More specifically, we can, and will, define as an indicator of patterned-cyclical time consciousness an interactional variable, a function of the *product* of the two variables (defined in the next chapter).

Up until this point, we have developed the theory of time and society, formally stated in the six propositions of the theory, developed a rationale for using life-historical interviews as data to study the theory empirically, and introduced the methodology, NHCA, to be used for this purpose. All that is left to do is describe the study, the corpus of interviews, the measurement of variables, the method of analysis, and the results of the analysis.

Chapter 15

An Empirical Test of the Theory

THE LIFE-HISTORICAL INTERVIEWS

The data set for this study consists of carefully edited transcripts from a corpus of 658 life-historical interviews, with 383 Aborigines (204 males and 179 females) and 275 Euro-Australians (155 males and 120 females). These interviews were obtained throughout Australia and are roughly representative of the two subpopulations. The collection of interviews is called a corpus rather than a sample, because random sampling was not used. Australia is a multicultural society by any measure, but the non-Aboriginal interviews were, in large measure, selected on the grounds that they trace their ancestry primarily to the British Isles and Northern Europe in an effort to reduce within-sample variation. For this reason, members of this portion of the corpus will be referred to as Euro-Australians. Many of the interviews were obtained by the author, in collaboration with Aborigines from the New South Wales Aboriginal Family Education Centres Federation, while others were obtained from institutes, libraries, private collections, and publications.

MEASUREMENT OF VARIABLES

In order to have some confidence that the words selected as indicators of folk concepts are not measuring different concepts, for each candidate folk concept an item analysis based on the method of summated ratings (see Edwards 1957: 149–57) was carried out for all of the selected words assigned to every Roget folk concept. This was done in two stages. In the first stage of item analysis, a summated rating—the proportion of total words spoken by the informant in the entire interview assigned to each folk concept—was calculated. The top quarter and bottom quarter of the sample were then compared, and then two-sample t-tests of differences between the means for the top and bottom quarters of the corpus were calculated separately for each word in the word list.[1] If an individual word measures what the words measure collectively, then the mean for the top quarter should be higher than the mean for the bottom quarter. In the second stage, the top half and bottom half of the corpus were compared, with t-values again calculated for each word. In both stages, if one or both of the two ts for a word was negative, then the word was purged; if one t was $\geqslant 1$ and the other was not computed (for rarely used words) or had a value between 0 and 1, then the word was retained; and, of course, if both t-values were $\geqslant 1$, then the word was retained. Two example word lists, for *Will* (an indicator of episodic-futural time consciousness) and *Demand* (an indicator of the positive experience of authority ranking) are shown in table 15.1. If a person in her or his cognitive structure emphasizes episodic information processing in general and has an episodic-futural time consciousness in particular, then she or he should be expected to make use of the words grouped under the Roget category *Will*, as they refer to the exercise of willpower in the words uttered in life stories. This, simply put, is the rationale for the content-analytic methodology.

Roget folk categories were selected, on theoretical grounds, as indicators for (1) the eight social-relations variables—the positive and negative experiences of communal sharing (*CS-pos, CS-neg*), market pricing (*MP-pos, MP-neg*), authority ranking (*AR-pos, AR-neg*), and equality matching (*EM-pos, EM-neg*); and (2), folk categories were selected to measure the four kinds of time consciousness—patterned-cyclical (*PC*), ordinary-linear (*OL*), episodic-futural (*EF*), and immediate-participatory (*IP*).

The twelve sets of items were subjected to a maximum likelihood factor analysis, and Tucker-Lewis (TL) inter-indicator reliability coefficients were calculated. The detailed results of these analyses are shown in the Appendix (table A.1). For these dozen variables, the final measure was the total number of words used from the list of folk-concept indicators, divided by the total words produced in the whole interview, with this quotient then multiplied by 10^4.

The ten *PC* and nine *CS* folk-category indicators span their seven-part definitions, as there is at least one measure for each of P1–P7 and L1–L7,

TABLE 15.1 Example word list for two roget folk categories, demand (indicating the positive experience of hierarchical ranking) and plan (indicating episodic-futural time consciousness); *t*-values compare top one-fourth and bottom one-fourth of proportions of words spoken belonging to the selected words

Demand

ask	6.96	demand	4.81	insistent	2.02	taxed	1.42
asked	8.96	demanded	3.72	insisting	1.57	taxes	1.00
asking	6.12	demanding	3.74	insists	1.50	taxing	1.42
asks	1.42	demands	3.47	levy	1.00	taxes	1.00
blackmail	1.42	direction	3.66	requisition	1.51	tribute	2.16
blackmailed	1.41	directions	3.61	requisitions	1.51	tributes	1.00
claim	5.25	duties	2.73	stipulated	2.02	ultimatum	1.00
claimed	2.83	duty	3.74	stipulation	1.00	urge	2.49
claiming	2.37	insist	2.39	superimposed	1.42	urged	2.80
claims	3.60	insisted	3.67	tax	2.84	urging	2.89

Will

choice	4.20	desire	3.65	initiatives	1.00	willing	3.78
choices	1.46	desired	1.07	solve	3.70	wills	3.38
choose	4.21	desires	1.46	solved	3.02	wish	5.67
chooses	2.02	desiring[a]	–	solving	2.26	wished	4.22
choosing	1.16	desirous[a]	–	volition	1.00	wishes	3.14
chose	5.05	fate	3.07	will	8.80	wishing	3.02
chosen	5.12	initiative	2.44	willed	2.03		

[a] These two words were not used enough to calculate *t*-values for the top and bottom quarters but are included because their *t*-values for the top and bottom halves were ⩾1. Source: TenHouten 2004: 22.

which contributes to their content validity. The second criterion for *PC* and *CS*, P2 and L2, were measured by interaction terms: P2, a fusion of the past and the present, was measured by *PastPres* = (Past ∗ Present)$^{1/2}$ and L2 by *FuturePres* = (Future ∗ Present)$^{1/2}$. For two of the *OL* criteria, word lists were pooled: for L1, time as linear, the measure used was *Linear* = Length + Interval + Period; and for L6, *Quantity* = Measurement of Time + Frequency. While *OL* is, by its seven-part definition, interpretable as a single dimension, *PC* is, by its definition, multidimensional, a difference that is reflected in the Tucker-Lewis reliabilities *OL* .90 and *PC* = .67, which suggests that *OL* came closer to being a single dimension than did *CS*.

Episodic-futural time was measured by seven Roget word lists that would appear to be face-valid and have good factor-score coefficients, but the TL reliability of .49 was the lowest of the twelve variables. *Immediate-participatory* time had only three indicators, so one, *The Present*, was randomly split into two variables to overidentify the model and make possible a reliability estimate (TL = .81), which had no effect on the combined measure of *IP*.

The four most important independent variables—the positive experiences of the four social relations—were adequately measured, as the TLs were *CS-pos* .98, *MP-pos* .99, *AR-pos* .77, and *EM-pos* .88. For *EM-pos*, a random split of the Roget indicator *Equality* was carried out to overidentify the measurement model and thereby make possible an assessment of reliability. The negative experiences of the four social relations were less well measured, as the TLs were *CS-neg* .61, *MP-neg* .83, *AR-neg* .91, and *EM-neg* .58. It should be mentioned that Roget provides many folk concepts that have an unclear or a neutral valence (especially for *MP*-related categories). For example, the potential *CS-pos* categories "Birth" and "Marriage" did not factor in with the chosen indicators and upon reflection cannot be assumed to be positive experiences, for the interviews contained many stories of tragic problems in giving birth, failed and abusive marriages, and having children "taken away": however, they were included indirectly in the measure *Temporality* = (Birth * Reproduction * Marriage)$^{1/2}$.

To explore the construct validity of the social-relations variables, the means of these variables were calculated for each of the four culture-sex groups. The data (not shown) are consistent with the literature and make common sense: (1) females were more verbally expressive of *CS* relations, both positive and negative, than males; (2) Euro-Australians were more involved in *MP* relations, both positive and negative, than Aborigines; (3) Aborigines expressed an insufficiency of social power (*AR-neg*) relative to Euro-Australians; and (4), Aborigines, relative to Euro-Australians, expressed a high level of denigrated identity (*EM-neg*).

The positive and negative experiences of hedonic society were defined by the following interaction terms: *Hedonic-pp* = (*CS-pos* * *EM-pos*)$^{1/2}$ and *Hedonic-nn* = (*CS-neg* * *EM-neg*)$^{1/2}$, respectively; the positive and negative experiences of agonic society, by *Agonic-pp* = (*MP-pos* * *AR-pos*)$^{1/2}$ and *Agonic-nn* = (*MP-neg* * *AR-neg*)$^{1/2}$. The four interaction terms, mixing positive and negative social relations and the two pairing negative relations, defined in a similar way, played no role in exploratory regression analyses and were excluded from the final analyses.

The natural time (*NT*) and rational time (*RT*) variables were operationally defined as *NT* = (*PC* * *EM*)$^{1/2}$ and *RT* = (*OL* * *EF*)$^{1/2}$, respectively. The four measures of hedonic and agonic social relations were used only in the two regression analyses using *NT* and *RT* as the dependent variables.

The six criterion variables, the measures of time consciousness, were then subjected to univariate analysis in order to normalize their distributions. Square-root transformations were necessary for *PC, EF,* and *IP*. After transformation, these measures were approximately normal, with the Shapiro-Wilk coefficients (1 = perfectly normal) *PC* .96, *OL* .98, *EF* .98, *IP* .97, *RT* .96, and *NT* .97.

A number of covariates and cofactors were defined. The informants' ecological location as a child and an adult was recorded, the settings being rural-outback, rural, small urban, and suburban-urban. Using s*mall-urban* as a reference category, three dichotomous (1, 0) variables were defined: *outback-tribal* = 1 if the informant lived in such a location both as a child and an adult, and 0 otherwise; *rural* = 1 for rural residence at both stages of life, 0 otherwise; and *urban* = 1 if location if suburban-urban at both life stages, 0 otherwise. The variable *Culture* was coded Aborigines 1 and Euro-Australians 0; sex, males 1 and females 0; the *Culture-Sex* interaction, *CS* = *Culture* * Sex. The covariates *Age in Years* and *Year of Birth* had missing values assigned to the mean for exploratory regression analyses, but the missing-values cases were excluded for these two variables in other analyses.

REGRESSION ANALYSES

The detailed results of the six ordinary least-squares regression analyses are shown in the Appendix (table A.2). The predicted standardized partial regression coefficient β values (which have t-distributions) and their associated probability ranges are shown in boldface along the main diagonal of the portion of table A.2 above the cofactors *Culture* and *Sex*. These six coefficients are predicted to be positive and in the range of statistical significance. The effects of social relations variables will be presented first, to test Propositions 1–6, and then the effects of the cofactors and two covariates will be described.

PROPOSITION 1

Patterned-cyclical time consciousness was regressed on the eight social relations variables (table A.2, column 1), *Culture*, and *Sex*. The prediction is that the positive experience of communal sharing, when all other variables are controlled, contributes to PC, and this is the obtained result (the standardized partial regression coefficient, $\beta = 3.36$, with a one-tailed probability, $p < .001$). *CS-neg* was unrelated to *PC*, and all of the other six social relations variables had negative coefficients, one of which (*MP-neg*), which is the opposite of *CS-pos*, was significantly negative.

Proposition 2

Ordinary-linear time consciousness was regressed on the eight social relations variables. As predicted, it was the positive experience of market pricing, *MP-pos* ($\beta = 2.85$, one-tailed $p < .01$), that contributed to this time orientation. The other component of agonic society, *AR-pos*, also made a significant positive contribution to *OL*, which is consistent with theory but not predicted. It also was found that *EM-pos* significantly depressed *OL*.

Recall Thompson's (1967) argument that a linearity of time consciousness has historically been imposed on working-class persons. Given this argument, we might expect a similar difference by *Culture*, as we have seen evidence that linear time is imposed on Aborigines working for Euro-Australians under oppressive conditions. To investigate this possibility, *OL* was regressed on the same independent variables separately by *Culture*. For the Euro-Australians, *MP-pos* again had a strong effect on *OL* ($\beta = 3.06$, one-tailed $p = .001$), but *MP-neg* was no longer unrelated to *OL* but instead depressed *OL* ($\beta = -2.84, p = .002$). But for Aborigines, *MP-pos* made only a directional and nonsignificant contribution to *OL* ($\beta = 0.98$, one-tailed $p = .16$), but *MP-neg*, the negative experience of market-pricing social relations, made a strong, positive, significant contribution to an ordinary-linear time consciousness ($\beta = 3.15$, two-tailed $p = .02$)! This remarkable result is entirely consistent with Thompson.

Proposition 3

Episodic-futural time consciousness was, as predicted, strongly and positively influenced by *AR-pos* ($\beta = 6.57$, one-tailed $p < .0001$), and also by the other positive component of agonic society, *MP-pos* ($\beta = 4.75$, $p < .0001$), with the other six coefficients, including *AR-neg*, at chance level.

Proposition 4

An immediate-participatory time consciousness was, as predicted, strongly and positively related only to *EM-pos* ($\beta = 9.21$, one-tailed $p < .0001$), but not to *EM-neg* ($\beta = .44$). The other six coefficients were negative, three of these results being statistically significant.

Proposition 5

For the *entire corpus* of interviews, natural time experience (*NT*) was, according to theory, expected to be predicted by the positive experience of hedonic social relations, *Hedonic-pp*, and this was obtained ($\beta = 2.51$, one-tailed $p < .01$). Other results, not predicted but consistent with theory, are of interest.

(1) *NT*, as predicted, responded positively to one component of hedonic sociality, the positive experience of equality matching (for *EM-pos*, $\beta = 2.81$, two-tailed $p < .01$), but not to the other component, the positive experience of communal sharing ($\beta = -1.32$, not statistically significant [n.s.]);

(2) *NT* was diminished at least directionally by all six of the other social-relational variables and significantly so by the negative components of agonic sociality (for *MP-neg*, $\beta = -2.95$, two-tailed $p < .01$; for *AR-neg*, $\beta = -2.99$, two-tailed $p < .01$). These results suggest that the experience of economic deprivation and an insufficiency of social power do not enhance informal social and family life but rather would appear to have a destructive effect, these results being consistent with a vast sociological literature documenting the tangle of pathology resulting from economic disadvantage and social powerlessness.

Separate analyses also were carried out for each *Culture*.

For the *Aborigines only*, the results parallel that of the entire sample. The theoretical prediction that *NT* would respond positively to *Hedonic-pp* was satisfied ($\beta = 2.00$, one-tailed $p = .02$). Again, there was a positive effect of hedonic component *EM-pos* ($\beta = 2.08$, two-tailed $p = .02$), but not for component *CS-pos* ($\beta = -1.46$ [n.s.]). And again, the other six variables had negative regression coefficients, with the same two nearly significant or significant (for *MP-neg*, $\beta = -1.46$, two-tailed $p = .14$; for *AR-neg*, $\beta = -3.23$, two-tailed $p = .001$).

For the *Euro-Australians only*, the results again resemble those of the entire sample, and Aborigines only, but they tend to be weak and not significant. *NT* responded only directionally to *Hedonic-pp* ($\beta = .79$, one-tailed $p = .21$). As before, *EM-pos* made a positive contribution ($\beta = 2.20$, one-tailed $p = .01$), but *CS-pos* did not. Economic difficulty, as indicated by *MP-neg*, again had a negative effect ($\beta = -3.10$, two-tailed $p = .002$), but a deficit of social power had only a weakly negative effect ($\beta = -.38$ [n.s.]). These interesting, and important, cultural differences will be discussed in the next chapter.

Proposition 6

Again, we begin with an analysis of the entire corpus. Rational time experience (RT) was, as proposed, predicted by the positive involvement in *agonic* society ($\beta = 4.37$, one-tailed $p < .0001$). One component of positive, agonic social relations, *AR-pos*, made an independent contribution to RT ($\beta = 2.45$, $p < .01$), but the other, *MP-pos*, had a nonsignificant negative coefficient.

Separate analyses by *Culture* were then carried out.

For the *Aborigines*, RT was successfully predicted by *Agonic-pp* ($\beta = 2.00$, one-tailed $p = .02$), with an independent contribution from *AR-pos* ($\beta = 2.55$, one-tailed $p < .01$). We again find a Thompson-type effect, as *MP-neg* has a significant effect ($\beta = 2.17$, two-tailed $p = .03$), but *MP-pos* had no effect whatsoever. There also was an unexpected negative effect of *EM-pos* ($\beta = -2.45$, two-tailed $p = .01$), suggesting that agonic and hedonic orientations might, to some extent, be a zero-sum in their effects.

For *Euro-Australians*, RT was strongly influenced by *Agonic-pp* ($\beta = 4.84$, one-tailed $p < .0001$), the inclusion of which reduced the effects of the agonic components to *negative* status, with the effect of *MP-pos* significantly negative ($\beta = -2.31$, one-tailed $p = .01$).

Effects of Culture, Sex, and Other Variables

The effects of *Culture* are consistent with the ethnographic literature on Aboriginal time consciousness and are very strong for all six kinds of time. As hypothesized, the Aborigines, in comparison to the Euro-Australians, were found to have a temporal orientation that is patterned-cyclical ($\beta = 7.18$, one-tailed $p < .0001$), present-oriented ($\beta = 6.61$, one-tailed $p < .0001$), and natural ($\beta = 9.00$, one-tailed $p < .0001$), which follows from their culture's great emphasis on community life, on equality-matching social relations, and on the combination of the two, which is constitutive of hedonic sociality. The Euro-Australians, in contrast and as hypothesized, give more emphasis to ordinary-linear time ($\beta = -7.21$, one-tailed $p = <.0001$), to the future ($\beta = -4.90$, one-tailed $p < .0001$), and to a rational time orientation ($\beta = -5.94$, one-tailed $p < .0001$), which follows from their involvement in market pricing, authority ranking, and the combination of the two, which comprises agonic society (see figure 15.1).

The effects of *Sex* were generally weaker than those for *Culture*, but all except one were statistically significant. Female informants showed a time

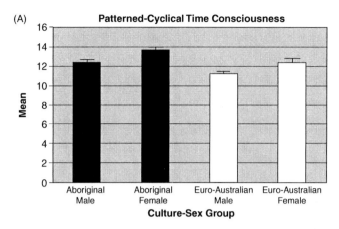
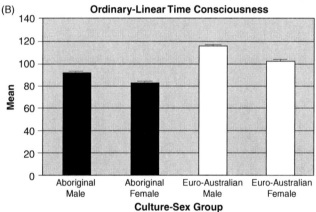
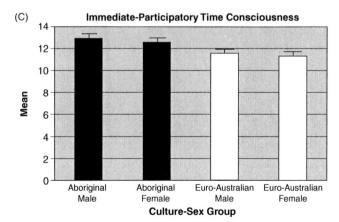

FIGURE 15.1 Means values of the six kinds of time consciousness, by culture and sex. (A) patterned-cyclical; (B) ordinary-linear; (C) immediate-participatory;

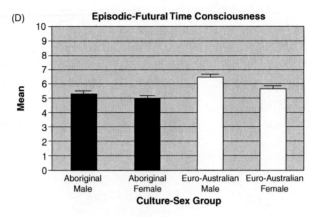
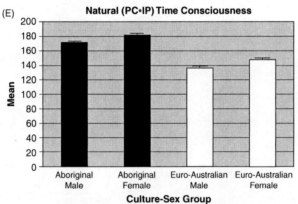
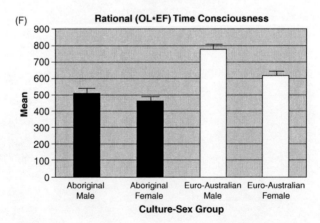

FIGURE 15.1 *(Contd)* (D) episodic-futural; (E) natural, the product of patterned-cyclical and immediate-participatory; (F) rational, the product of ordinary-linear and episodic-futural. Error bars are + standard deviation.
Source: TenHouten 2004: 28.

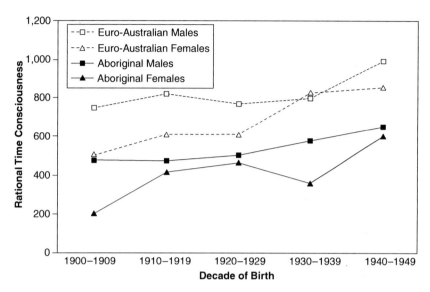

FIGURE 15.2 Mean rational time consciousness as a function of *Decade of Birth*.
Source: TenHouten 2004: 30.

consciousness that was strongly patterned-cyclical ($\beta = -.72$, one-tailed $p < .0001$) but slightly less present-oriented ($\beta = 2.90$, two-tailed $p = .003$, a result not predicted), the net effect being a slightly lower level of natural time ($\beta = 1.80$, two-tailed $p = .07$). Males, in contrast, were much more linear ($\beta = 4.00$, one-tailed $p < .0001$) and future-oriented ($\beta = 2.48, p < .01$), and higher for rational time ($\beta = 3.35, p < .001$).

The regression analyses presented in the Appendix (table A.2) were first done with the *Culture-by-Sex* interaction included, but none of these interaction terms was significant, so this variable was returned to residual status before the final analyses were carried out.

The above analyses were repeated with the inclusion of other variables, which resulted in some missing data but the same overall results (data not shown). There was an interesting difference for *Age*: the older informants were more present-oriented but less future-oriented than the younger ones.

The variable *Year of Birth* makes possible inferences about historical trends in the mentalities of the two cultural groups. The data show that as the decades (from the 1900s through the 1950s) pass, there has been a modest increase in linearity and in futurality, and a strong increase in rational time. These trends can be seen in figure 15.2, where some interesting culture-sex differences are revealed. This trend is strongest among the Euro-Australians (their line "slopes" are steeper), with the females appearing

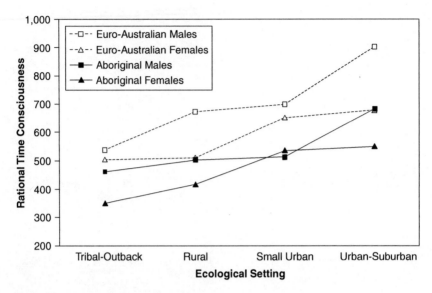

FIGURE 15.3 Mean rational time consciousness as a function of *Ecological Setting*. *Source*: TenHouten 2004: 29.

to be closing the gap for informants born in the 1940s and 1950s. Among the Aborigines, in contrast, the sex difference has not narrowed. The overall tendency toward a rational time orientation with the development of modern, industrial society is well documented in the literature on time and temporality (see, e.g., Heidegger [1927] 1962; Whitrow 1989). Natural time, in contrast, has been stable over the decades, and least for these data (results not shown).

Rational time also is strongly linked to ecological setting, as can be seen in figure 15.3. As might be expected, it is the outback, "bush"-dwelling Euro-Australians and the tribal-living Aborigines who have the lowest levels of rational time, with monotonic increases evident in larger settings, in rural, small urban, and then in urban-suburban locations. The effect of location appears stronger for males than females in both cultural groups.

Chapter 16

Discussion

DISCUSSION OF RESULTS OF THE STUDY

The present theory of time and society is based on three isomorphic models, one of social relations and two of cognitive structure. The four elementary forms of temporality identified in the cognitive model are seen as aspects of four larger cognitive structures, the four most major modes of information processing of the human brain. Once these models are developed, the theory itself is rather simple, and in this first empirical study of the theory, the fit to data is excellent. As predicted, patterned-cyclical time consciousness is predicted by the positive experience of communal-sharing social relationships; ordinary-linear time consciousness by the positive experience of market pricing; episodic-futural time consciousness by the positive experience of authority ranking; and immediate-participatory time consciousness by the positive experience of equality matching.

The negative experiences of these four kinds of social relations produced only five out of twenty-four possible significant results in the whole-corpus analyses, all of which had negative coefficients. First, both *patterned-cyclical* and *natural* kinds of time consciousness were significantly lowered by the negative experience of market-pricing, economic relationships. And second, a present orientation was significantly lowered by the negative experiences of CS, MP, and AR: it should be noted that the *positive* experiences of CS, MP, and AR also depressed immediate-participatory time consciousness (*IP*). None of these negative coefficients is of particular interest to the theory. What is important is the predicted result, the strong, direct relationship between *IP* and *AR-pos* that far exceeded the other social-variable results in its strength and level of improbability ($\beta = 9.22, p < .0001$).

Recall that Fiske (1991) has been criticized for not paying attention to the valences of his four social relations models. The importance of doing so is evident from these findings. For the entire sample, it was only the *positive* experiences of the four social relations that contributed positively to the four kinds of time consciousness, and it was only the *products* of the *positive* pairs of social relations that predicted rational time (RT) and natural time (NT) experiences. The only negative social relationship that predicted a kind of time consciousness that was of theoretical interest was found in the analysis of Aborigines only, where it was found, consistent with Thompson (1967), that negative experience of the work world contributed to a linearity of time consciousness. Further analysis, with the addition of a measure of socioeconomic status, is of course needed to better understand this phenomenon.

The importance of the most general kinds of time experience, the natural and the rational, is clearly evident in the results.

Natural time is predicted by the interaction term of positive hedonic community, but when this variable is added to a regression analysis based on the eight social relations, only one of its two components—the positive experience of equality matching, but not by the positive experience of communal sharing—continues to have an effect.

Rational time is predicted by the interaction term of positive agonic sociality, but when this variable is added to a regression analysis based on the eight social relations, again we find that only one component—the positive experience of authority ranking, continues to have an effect. The conclusion is that the higher-level concepts, natural-rational, and hedonic-agonic, are hardly supererogatory to the theory but rather are essential. It cannot be said that the positive experience of communal sharing has no effect on NT, nor can it be said that the positive effect of market pricing has no effect on RT. What can be said is that the effects of *CS-pos* and *MP-pos* are not direct but rather come about in their interactions with their partners, *EM-pos* and *AR-pos*, respectively.

While it is claimed that the four social relationships—CS, MP, AR, and EM—are cultural universals, no such claim has been made for the two higher-order concepts, hedonic and agonic society. Recall that propositions 5 and 6 are of an if-then nature, stating "to the extent that cultural members participate" in positive hedonic or agonic society, their time consciousness should be natural or rational. The data suggest that Aborigines have a time consciousness that in addition to being patterned-cyclical and present-oriented, is also natural, and the concept of natural time is given *predictive validity* by its responsiveness to positive hedonic sociality. The data also show that Euro-Australian time consciousness, in addition to being linear and future-oriented, is also rational, with the concept of rational time given predictive validity by its responsiveness to positive agonic sociality.

Whether there would be evidence that Aboriginal culture also has developed an agonic aspect as a component of its level of cognitive assimilation, or cognitive adaptation, was at the beginning of data analysis an open question. What was found was that the measure of positive agonic sociality *did* predict rational time, which provides predictive validity for the existence of agonic sociality among the Aborigines. But we did not find significant evidence that Euro-Australian culture is hedonic, because the hedonic sociality measure for it *did not* significantly predict natural time. This, of course, does not mean that a hedonic community is entirely absent among Euro-Australians, for there is abundant evidence that such a community does exist, including the author's personal observations of Euro-Australian family and informal social life.

The finding that rational time experience has increased over decade of birth (figure 15.2), whereas natural time experience has been relatively stable, while not part of the formal theory, is important. Linear, episodic, and rational kinds of time consciousness have, during the twentieth century, increased historically (increasing as a function of decade of birth), while cyclical, present-oriented, and natural time consciousness have not changed over decade of birth. Since Weber ([1922] 1978), it has been widely believed that modern societies have undergone a period of progressive rationalization, and a progressive rationalization of time consciousness can be considered part and parcel of this larger process. It certainly should not be inferred from the data presented that the progressive rationalization of time consciousness has rendered a natural time consciousness any less important to the overall effective functioning and adaptation of the human mind in its sociocultural context. In fact, it is very possible that a *pathological* lack of a natural time consciousness is part and parcel of a mentality that sees the earth and the life on it not as a fragile, delicately balanced web of life but rather as resources to be developed, exploited, and consumed, at the cost of the degradation and destruction of oral, indigenous cultures and of vast ecological damage to the entire planet and its life.

On the level of methodology, an effort has been made to show the utility of quantitatively analyzing texts for the purpose of extracting information about both social relations and cognitive structure. The word-classification method developed for this study, NHCA, is neurocognitive insofar as the concepts used to measure mental proclivities are backed up by neuroscientific theory and concepts. It is hierarchical because it makes use of a hierarchical classification of the English language (classes, notions, categories, folk concepts, and words). Other classifications, of course, might be more appropriate if texts are studied with other cognitive structures in mind, or for other purposes altogether.

General Discussion

In her persuasive rejection of temporal dualism, Adam (1990: 16–19) refutes disciplinary isolation in the study of time and society. The present theory and research also refute disciplinary isolation, as it is shown that a multilevel, multidisciplinary approach is required for an understanding of time and society, which models the sociocultural, the mental, and the biological as three necessary levels of analysis. The growing interest in time in sociology and related social and behavioral scientific disciplines is an expression of the growing appreciation of mind and society as *the* unifying topic of social theory. Cognitive sociologists, such as Zerubavel (1997: 3), have urged that in this effort we steer a middle course between cognitive universalism, on the one hand, and local knowledge and cognitive individualism, on the other hand. But a theory of time and society *can* be constructed directly from the most modal sociocultural and cognitive universals, the elementary forms of sociality and the elementary forms of time consciousness, which, it is proposed, require that these concepts be criterion validated by showing that they have a biological basis and an evolutionary history.

Much comparative research on culture and cognition has been carried out in psychology and has made extensive use of psychometric testing. The limitations of cross-cultural, cross-societal comparisons based on psychometric tests standardized on the norms of Western, modern society are well known. Comparisons of mentality and cognitive ability based on such tests are widely viewed as discriminating against members of "primitive" or other nonmodern societies (and to subdominant groups and classes in modern societies), which generally are outperformed by their comparatively "modern" and/or "advantaged" controls and on this basis have historically been invidiously stereotyped as lacking intelligence. This is socially important because life chances will long remain linked to test performances. From a scientific point of view, such comparisons provide limited information about the overall mentalities of people who live in different societies and cultures. Another level of analysis is required for a comparative understanding of human mentality, of the human mind in its sociocultural context. It is the author's conviction that the one of the deepest and most fundamental levels of analysis that can possibly be used for comparative, historical, and cross-cultural analysis of mind and society is that of the relationship between sociocultural experience and time consciousness. Thus rather than rely on tests and measures, or focus on mind in general, a general theory of culture and time consciousness is presented, which, it is hoped, can stimulate further research and other cross-cultural comparisons.

The theory of time consciousness presented in this book is an original formulation. It should be mentioned, however, that a similar model of time consciousness has been proposed by Mann, Siegler, and Osmond (1968). In their Jungian framework, they deny that cultural and social ordering influences time consciousness, rather, arguing that time is a wholly private perception. Jung's (1923) system postulates four functions of the mind (and two attitudinal types, extraversion and introversion): sensation tells us that something exists, thinking, what that something is; feeling enables a value judgment about the object; and intuition enables us to see where it came from and where it is going. Corresponding to these four mental functions are four temporal orientations that vary in importance from person to person. The sensing type of person relates to the present, which is here called "immediate-participatory time consciousness"; the thinking type, to the time line, here called "ordinary-linear time consciousness"; the feeling type, to the past in relation to the present, one aspect of which is here called "patterned-cyclical time consciousness"; and the intuitive type, to the future, which roughly corresponds to the present notion of "episodic-futural time consciousness." Of the four kinds of time in this psychotypology, "relating to the past" fits the present model most poorly. Mann, Siegler, and Osmond see this kind of time as continuous, where the present model stresses its discontinuity. Moreover, the basis of patterned cyclicity is not "feeling" but gestalt synthesis, even though the right hemisphere is involved in the cognitive representation of feelings due in large measure to its central role in complex pattern discrimination. The theory presented here and the data in support of this theory see time consciousness not as a purely private matter of personality temperament, of a personal *Umwelt*, but as socioculturally influenced.

Any theory linking time consciousness to society must have as its basis some conceptualization of social organization. It has been argued that there are two basic kinds of societal organization, the *agonic* and the *hedonic*. As was shown in chapter 8, the elementary social relations come in two pairs, AR ∩ MP for agonic society and EM ∩ CS for hedonic society. Without the agonic-hedonic distinction, there would be a logical contradiction in the theory, because the primary emotion associated with market pricing is expectation/anticipation, which suggests that the positive experience of MP should contribute both to linearity and to futurality, which has been found to be the case.

The culture of the Australian Aborigines has been described as placing a great emphasis on equality, or making things "level." When an Aborigine does a favor for another Aborigine, there is no felt need for a "thank-you," because it can be assumed that the favor will be returned in an appropriate way in an appropriate span of time. The culture also has been described as placing a great emphasis on family, community, and tribe. More specifically, Aboriginal society has been described as emphasizing equality matching and communal sharing in social relationships. Thus we can refer to traditional Aboriginal culture as "hedonic," and to its kind of time consciousness as "natural."

It has been shown in chapters 10 and 11 that Aboriginal society emphasizes communal-sharing and equality-matching social relationships, and that Euro-Australian society places less emphasis on both. Then, in chapters 12 and 13, it was shown that Aboriginal society and culture have only limited involvement in authority-ranking and market-pricing social relationships, with Euro-Australian and, more generally, modern society and culture placing far greater emphasis on these social relationships. Because CS and EM are the complementary components of hedonic society, and AR and MP are the complementary components of agonic society, Aboriginal society is largely hedonic, and Western society is largely agonic.

The mix between the elementary forms of sociality, and the emphasis on agonic or hedonic society, varies both between and within cultures. The case study of a hunting-and-gathering society, the Australian Aborigines, emerges as a nearly *pure form* of hedonic society, with a great emphasis, even an insistent demand, on conditional equality, on making things "level," and also on communal sharing, while at the same time institutionalizing neither polity (authority ranking) nor a cash economy (market pricing), thus not developing an agonic society. A modern, Western society such as Australia's is a radical contrast. What best characterizes modernization is the extraordinary elaboration of *political economy*, the institutional *unification* of the social relations authority ranking and market pricing, a fundamental linking of power and money.

This argument most certainly does not mean, however, that Aboriginal culture, even on its most traditional level, excludes all agonic social relations, and that Euro-Australian culture excludes all hedonic social relations.

Australia's Aborigines of course do not live in isolation, and for them the modern, agonic society of modern Australia is an overwhelming reality that they must face every day of their lives. With this in mind, let us consider a story told by Uncle Bill, an eighty-year-old Aboriginal man from Old Burnt Bridge on the mid-north coast of New South Wales. What follows is a verbatim transcript from his life-history interview, with words spoken by the two interviewers (W. T. and K. C.) removed. The text is punctuated by ideas, with shifts of topic indicated by a blank line. Uncle Bill said the following:

I've traveled this river
up and down,
everywhere.
Now, we got up here, fishing.

Where we used to fish
along the bank,
we got notices back up now:
"Trespassers will be prosecuted."

Yeah,
they won't allow you in there
Oh, me and another chap
a terrible argument
up there,

I was digging a few worms.
And he come up to me
and he said
"What are you doing?"
And …
"digging a few worms for myself."
"Oh," he said,
"Well you'll have to move."

"Well," I said,
"Well, I'm not moving, man."
"I'm digging these worms here."

He said,
"I'll ring the police up."
Well I said,
"You'll ring the police up,
And I'll still be here."
I said,
mildly explained to him,
Why I mean,
they never come,
Yeah.

The police never come.
He just tried to frighten me off,
I mean, Yeah,
things went all right then,
I went down
and caught my fish
and throwed my line up
and walked off.

Ah, there's a lot of us
nearly all, everybody
gathering up along here
fishing.
Nearly all our people,
All along the riverbanks,
I've traveled the river up and down.

Agonic society is the unity of authority ranking and market pricing. This is crystallized in a collocation that is highly oppressive to Aborigines: "Trespassers will be prosecuted." Uncle Bill is confronted with the realities of the conquest and dispossession of his people. He is told that he no longer has access to his traditional land, to the river where his people have always lived. His territory has been expropriated, which is a negative experience of market pricing, market pricing being conceptualized here as a social extension of territoriality. He is now a trespasser, threatened with the police, which is a negative experience of authority ranking. Also evident in the story is Uncle Bill's courage. His aim is to participate in his culture's primordial mode of economic production, hunting and gathering. Uncle Bill is intent on fish hunting and to the preservation of his community's way of life along the riverbanks. He is intent on participation in his own hedonic culture, but in order to do so, he has had to cope with an agonic culture and society, and to this end, he has stood up to the intimidation of the Euro-Australian, which is a *positive* experience of authority ranking, and he has insisted on his right to dig for worms along the river, to fish the river, and this is a *positive*, assertive experience of territoriality, more generally of market pricing.

The Aborigines have to deal every day with the existential problems of territory and authority. Nothing is more important to their cultural survival than access to their traditional lands so that they can participate in a collectively represented meaning system that features a totemic landscape. While they trade religious items and other things as well, they did not in their traditional culture develop money. As for authority, they traditionally have their law, and they invest decision making for the group with tribal elders. They did not develop the institutions of tribal chiefs or headmen, so they essentially possessed no political system. In general, Aboriginal culture can be described as primarily hedonic and secondarily agonic. The contrasted Euro-Australian, Western culture, of course, also includes the principles of equality and community. Australia is a democracy that by law is compelled to treat its citizens equally, and the family remains the basic institution of society. But modern society has undergone a great institutional elaboration of polity and economy, expressed in forms of sociality that emphasize authority-ranking and market-pricing social relations. We saw in chapter 2 that Aborigines have been invidiously stereotyped as locked in a time warp, unable to cope with the modern world. But cope they must, and cope they do, and in the process, they must undergo a process of cognitive development, which in the example of Uncle Bill's story is a manifestation not of weakness but rather of strength and resolve, and the involvement with agonic society that he has courageously faced has expanded his mind so that he must master the clock and the calendar and concern himself with his personal future and the future of his people. It would be a sad state of affairs if Aborigines had not developed the cognitive flexibility to not only become, as Swain suggests, a people with two laws but also a people with two kinds of time orientation, the natural and the rational. Thus we can conceptualize Uncle Bill's mentality not as a passive, dependent cognitive assimilation but rather as an active, independent cognitive adaptation.

The model of four kinds of time consciousness developed in this book is based on two models of brain functioning, Pribram's (1981) and Luria's (1973) idea that the posterior cortex and the frontal cortex are involved in participatory and episodic processing, and on the neuroscientific theory of hemispheric specialization. Of these two models, it is the latter, lateralization theory, which is highly controversial. The idea that the left and right hemispheres are specialized for two complementary models of thought is far from universally accepted and has been subjected to incisive criticism (e.g., by Springer and Deutsch 1981). One problem is that this theory has been subjected to some rather wild speculation in what Bogen has called "the wild blue yonder" (e.g., Jaynes 1977, Sagen 1977), which the author has dealt with elsewhere (TenHouten 1992b). The idea that cultural differences exist in hemisphericity also has been dismissed by many neuroscientists (e.g., Springer and Deutsch 1981: 188–190), even though the field of social neuroscience is now well established (see, e.g., Cacioppo et al. 2002). In this book, a great deal of effort has been made to spell out what it is about culture that enables us to distinguish one culture from another in a manner that is consistent with not only neuroscience but also with ethology, all in an effort to criterion validate the key concepts of the theory. This was accomplished by a generalization of the four-problems-of-life model of Plutchik's and Maclean's into the four social-relations model of Scheler's and Fiske's.

It has been argued here that there are universal structures of human social organization and that there are universal structures of the human mind and human brain—which include four kinds of time consciousness existing as aspects of four more general kinds of information processing. However, it is not the case that this work can be classified as any sort of structuralism, which it is most certainly not, or as ahistorical, which it also is not (on the history of time, see, e.g., Whitrow 1989; Terdiman 1993; Friedland and Boden 1994; Dark 1998). Instead I have argued that the human being has the capability for both hedonic and agonic society, for both natural and rational time consciousness, and the responsibility to develop healthy and productive social arrangements that are both hedonic and agonic, and that in striving for a rational orientation, we do not lose sight of the natural, for to do so would be, and has been, a gross violation of human responsibility to make a better world, one in which peoples with differing mentalities can understand that in spite of their differences, they have a common responsibility to respect and celebrate both their differences and their commonalities.

Appendix

TABLE A.1 Roget folk-concept indicators of the four time consciousness and eight social-relational variables, factor score coefficients, Tucker-Lewis reliability coefficients (TL), and total number of words (NW) for each Variable

Time Consciousness Variables

Patterned-Cyclical		Ordinary-Linear		Episodic-Futural		Immediate-Participatory	
Duality (P1)	.30	Linear[a] (L1)	.52	Will	.34	The Present-a[d]	.75
Past-Pres (P2)	.16	Fut-Pres[b] (L2)	.36	Resolution	.40	The Present-b[d]	.13
Infrequency (P3)	.10	Regular (L3)	.39	Intent	.56	Presence	.12
Season (P4)	.06	Age Years (L4)	.36	Plan	.54	Imminence	.34
Youngster (P5)	.37	Timeliness (L4)	.33	Foresight	.43	(TL .81, NW 47)	
Adult (P5)	.20	Earliness (L5)	.22	Presentiment	.22		
Interim (P6)	.23	Sequel (L5)	.14	The Future	.27		
Durability (P7)	.57	Quantity[c] (L6)	.37	(TL .49, NW 285)			
Evening (P7)	.30	Transience (L7)	.17				
Lateness (P7)	.32	(TL .90, NW 294)					
(TL .67, NW 297)							

Social-Relations Variables

Communal-Sharing-pos		Market-Pricing-pos		Authority Ranking-pos		Equality-Matching-pos	
Lovemaking	.21	Possessor	.13	Master	.05	Identity	.23
Friends	.50	Possession	.26	Demand	.33	Equality-a[g]	.84
Temporality[f]	.32	Acquisition	.91	Compulsion	.26	Equality-b[g]	.31
Welcome-Frnshp[e]	.34	Property	.16	Strictness	.40	Similarity	.10
Cooperation	.004	Wealth	.10	Disobedience	.28	(TL .88, NW 42)	
(TL .98, NW 297)		(TL .99, NW 120)		Opposition	.29		
				Resistance	.22		
				Contradiction	.31		
				(TL .77, NW 272)			

(*Continued*)

TABLE A.1 (*Continued*)

Social-Relations Variables

Communal-Sharing-neg		Market-Pricing-neg		Authority-Ranking-neg		Equality-Matching-neg	
Death	.32	Loss	.26	Lack Influence	.07	Contrariety	.18
Divorce	.08	Relinquish	.98	Confinement	.90	Difference	.08
Seclusion	.39	Poverty	.49	Obey	.12	Disapproval	.34
Selfishness	.29	Debt	.03	Prohibit	.07	Disagreement	.05
Dislike	.42	Payment	.23	Accuse	.73	Disrepute	.11
Discourtesy	.16	(TL .83, NW 138)		Condemnation	.15	Disparagement	.73
(TL .61, NW 118)				Punishment	.09	Ridicule	.04
				Atonement	.08	Injustice	.15
				(TL .91, NW 255)		(TL .58, NW 242)	

[a] Linear = Length + Interval + Period
[b] Fut-Pres = Future × Present
[c] Quantity = Measurement of Time + Frequency
[d] To overidentify the model, necessary for estimating inter-item reliability, the word list for "The Present" was randomly divided into two sublists
[e] Welcome-Frnshp = Welcome × Friendship
[f] Temporality = Birth × Reproduction × Marriage
[g] To overidentify the model, the word list for the folk concept "Equality" was randomly divided into two sublists.
Source: TenHouten 2004: 24.

TABLE A.2 Ordinary least-squares regression analyses for the combined samples: patterned-cyclical, ordinary-linear, episodic-futural, immediate-participatory, natural, and rational kinds of time consciousness regressed on social relations variables, culture, and sex (predicted βs in boldface)

Independent Variable	Patterned-Cyclical		Ordinary-Linear		Episodic-Futural		Immediate-Participatory		Natural = (PC × IP)		Rational = (OL × EF)	
Communal Sharing-pos	.016 (.005)	**3.36*****	.089 (.080)	1.11	−.005 (.005)	−1.00	−.004 (.003)	−.73	−.187 (.141)	−1.32	−.006 (.889)	−.07
Communal Sharing-neg	−.000 (.006)	0.01	−.054 (.099)	−.54	−.014 (.006)	−2.21*	−.014 (.006)	**−2.226***	−.206 (.117)	−1.76	−.064 (.074)	.87
Market Pricing-pos	−.004 (.008)	**−2.03***	.392 (.138)	2.85**	.039 (.008)	4.75***	−.042 (.009)	**−4.89*****	−.344 (.326)	−1.05	−.227 (.205)	−1.11
Market Pricing-neg	−.008 (.001)	**−4.60*****	−.027 (.099)	−.27	.010 (.006)	1.72	−.016 (.006)	**−2.48***	−.347 (.118)	−2.95**	.071 (.074)	−.96
Authority Ranking-pos	−.003 (.003)	−1.06	.085 (.088)	−.97	.018 (.003)	**6.57*****	.001 (.003)	−.24	−.042 (.063)	−.66	.098 (.040)	2.45*
Authority Ranking-neg	−.009 (.005)	−1.70	−.085 (.088)	−.97	−.004 (.005)	−.72	−.020 (.006)	**−3.64*****	−.313 (.105)	−2.99**	−.055 (.066)	−.83
Equality Matching-pos	−.001 (.001)	−.53	−.080 (.022)	−3.69***	−.006 (.001)	−4.47***	.013 (.001)	**9.22*****	−.093 (.033)	2.81*	−.069 (.021)	−3.31**
Equality Matching-neg	−.014 (.008)	**−1.82***	−.077 (.127)	−.61	−.006 (.008)	.73	.004 (.008)	.44	−.184 (.153)	−1.21	.019 (.096)	−.20
Hedonic-pp = (CS-pos*IM-pos)	—		—		—		—		.289 (.115)	2.41**	−.033 (.072)	−.44
Agonic-pp = (MP-pos*HR-pos)	—		—		—		—		−.260 (.254)	−1.03	.696 (.159)	4.37***
Culture	1.26 (.175)	7.18***	−20.65 (2.86)	−7.21***	−.847 (.173)	−4.90***	1.20 (.181)	6.61***	31.3 (3.474)	9.00***	−12.97 (2.183)	−3.35***
Sex	−.999 (.170)	−5.72***	11.420 (2.852)	4.01***	.427 (.172)	2.48*	.524 (.181)	2.90**	−6.08 (3.383)	−1.80	7.12 (2.125)	3.35***
R^2	.190		.149		.198		.272		.251		.229	

* = $p \leq .05$, ** = $p \leq .01$, *** = $p \leq .001$ (All of the one-tailed tests are in boldface.)

Note: Unstandardized coefficients are to the left of the standardized β-values, with these coefficients' standard errors directly below them in parentheses.

Source: TenHouten 2004: 26.

Notes

Chapter 2

1. Europeans battled their way into what they defined as "empty country," *terra nullius*, and then, in many cases, killed its inhabitants as a means of gaining control (Reynolds 1974).

2. Aboriginal activist Cheryl Buchanan (1988: 260–61) relates the story of what happened to one Aboriginal woman in Brisbane: "The Ship Inn is a hotel where blacks drink. There's a park next to it. A black woman was raped by a white man, and she went to the police. She knew the bloke; he drank regularly at the Ship Inn. She asked the police if she could press charges against him. ... This white bloke found out she'd asked the pigs, and when she went back to the pub he got her when she was walking past the park, raped her again, and then broke a bottle and ripped her vagina completely, right up to her backside. ... There's case after case of black women getting raped. You know if you're going to go to the pigs that either they'll keep you there and rape you themselves. I mean, this is just a known fact in the black community."

3. This estimate includes fifty extinct languages, 130 with less than fifty speakers but in limited use or still known to older generations, forty-five with fifty to 150 speakers, and twenty-five with more than 250 speakers (Yallop 1982: 30). Recent comparative studies carried out in regions of Australia (Dixon 1966; Sommer 1969) suggest that even the most aberrant Australian languages (excluding those of Tasmania) might well be variations on a more general regional type, an ancestral proto-language.

4. This putative family of Aboriginal languages shows no affinity for other language families of Southeast Asia or the Pacific. The main Austronesian language family of the Pacific includes some language of Papua New Guinea and Melanesia, as well as the Polynesian languages (including Maori). Some Aboriginal tribes of Arnhem Land, along the northeast Australian coastline, have borrowed from the Austronesian Macassans who used to visit Northern Australia. The ancestral development of the Australian languages is lost in the past and cannot be traced. The evidence at hand can support no assertion that there existed a single Australian proto-language. However, Yallop (1982: 33) is able to at least explore this topic, in the process providing a lexicostatistical analysis to infer time depths, that is, a "glottochronology" (see Gudschinsky 1956 for an explanation).

Chapter 3

1. "Mobs" are collections of Aboriginal people brought together in places where social purpose is not fixed for the long term but shifts instead in time (Sansom 1980: 14).

Chapter 5

1. According to Plutchik ([1962] 1991), territoriality/spatiality is one of four basic problems of life, and surprise is the adaptive reaction (primary emotion) to the negative experience of territoriality.

2. Neils Bohr discovered that electrons leap from one orbit to another in abrupt, inexplicable, unpredictable transitions called "quantum leaps," with their "location" only probabilistically distributed. This departure from Newton's view of a continuous world completed the overthrow, for subatomic phenomena, of the classical notion of causality (Slife 1981: 41).

3. Einstein's critique of Newton's absolute meaning of "simultaneity" undermined the notion of unidirectional time that was paradigmatic in classical physics. The cyclical-linear debate in physics was resolved by showing the relativity of simultaneity to the position of the observer. Relativity theory also showed the limitation of Newton's unidirectional view of causality. This revolutionary change in the conceptualization of time has had a limited effect on the behavioral and social sciences.

Chapter 10

1. In the analysis of data for this study, it was discovered that two cognitive processes are at work in visual closure tests: *pattern* recognition and *part* recognition, a process that involved recognizing individual parts of the pattern to provide clues to the whole.

2. A near-exception was discovered by the ancient Egyptians, who, using both of their lunar and 365-day calendars, found that 309 of their lunar months were almost equal to twenty-five of their civil years. The Egyptian 365-day calendar was used by Hellenistic astronomers and astronomers during the Middle Ages in Europe, for example, by Copernicus in his lunar and planetary tables (Whitrow 1989: 27).

Chapter 15

1. $t = (\bar{X}_{top} - \bar{X}_{bottom})/(s^2_{top}/n_{top} + s^2_{bottom}/n_{bottom})^{1/2}$.

References

Abram, D. 1996. *The Spell of the Sensuous: Perception and Language in a More-Than-Human World*. New York: Vintage Books.

Adam, B. 1990. *Time and Social Theory*. Cambridge: Polity Press.

———. 1995. *Timewatch: The Social Analysis of Time*. Cambridge: Polity Press.

Ahern, C. 1991. "Turkey Creek." Pp. 34–41 in *Aboriginal Art and Spirituality*, ed. R. Crumlin. North Blackburn, Victoria: Collins Dove.

Altom, M. W., and J. Weil. 1977. "Young Children's Use of Temporal and Spatial Order Information in Short-Term Memory." *Journal of Experimental Child Psychology* 24: 147–63.

Alverson, H. 1994. *Semantics and Experience: Universal Metaphors of Time in English, Mandarin, Hindi, and Sesotho*. Baltimore: Johns Hopkins University Press.

Ariotti, P. E. 1975. "The Concept of Time in Western Antiquity." Pp. 69–80 in *The Study of Time II*, ed. J. T. Frazer and N. Lawrence. Berlin: Springer-Verlag.

Aristotle. 1936. *Aristotle's Physics*. Introduction and commentary by W. D. Ross. Oxford: Clarendon Press.

Armstrong, E. 1982. "Mosiac Evolution in the Primate Brain: Differences and Similarities in the Hominid Thalamus." Pp. 131–61 in *Primate Brain Evolution*, ed. E. Armstrong and D. Falk. New York: Plenum Press.

———. 1991. "The Limbic System and Culture: An Allometric Analysis of the Neocortex and Limbic Nuclei." *Human Nature* 2: 117–36.

Axelrod, R. 1984. *The Evolution of Cooperation*. New York: Basic Books.

Bakan, P. 1966. *The Duality of Human Existence: An Essay on Psychology and Religion*. Chicago: Rand McNally.

Banks, R. 1983. "Attitudes to Time—Australian Aboriginal and Western." *Hemisphere* 27: 222–26.

Barbalet, J. M. 1998. *Emotion, Social Theory, and Social Structure: A Macrosociological Approach*. New York: Cambridge University Press.

Barden, G. 1973. "Reflections of Time." *The Human Context* 5: 331–44.

Barnes, R. 1974. *Kédang: A Study of the Collective Thought of an Eastern Indonesia People*. Oxford: Clarendon Press.

Beckett, J. R. 1965. "Kinship Mobility and Community among Part-Aborigines in Rural Australia." *International Journal of Comparative Sociology* 6: 7–23.

Beckett, J. R., ed. 1988. *Past and Present: The Construction of Aboriginality*. Canberra, Australian Capital Territory: Aboriginal Studies Press.

Bentham, J. 1812. *Panopticon versus New South Wales, or, The Panopticon Penitentiary System and Penal Colonization System Compared*. London: Wilks and Taylor.

Berdyaev, N. 1938. *Solitude and Society*. London: Centenary Press.

Bergland, R. 1985. *The Fabric of Mind*. Harmondsworth, Middlesex, England: Penguin.

Bergson, H. [1910] 1960. *Time and Free Will*. Translated by F. L. Pogson. New York: Harper.

———. [1911] 1975. *Creative Evolution*. Translated by A. Mitchell. Westpoint, Conn.: Greenwood Press.

Berndt, C. H., and R. M. Berndt. 1978. *Pioneers and Settlers: The Aboriginal Australians*. Melbourne: Pitnam.

Berndt, R. M., and C. H. Berndt. 1942–45. "A Preliminary Report of Field Work in the Ooldea Region, Western South Australia." *Oceania Bound Offprint*, 1945, Sydney (*Oceania* 12: 4; 13: 1–4; 14: 1–4; 15: 1–3).

———. 1948. "Sacred Figures of Ancestral Beings of Arnhem Land." *Oceania* 18: 309–26.

———. 1980. *Aborigines of the West: Their Past and Their Present*. Perth: University of Western Australia Press.

Bever, T. G., R. R. Hurtig, and A. B. Handel. 1976. "Analytic Processing Elicits Right Ear Superiority in Monaurally Presented Speech." *Neuropsychologia* 14: 175–81.

Blau, P. M. 1964. *Exchange and Power in Social Life*. New York: Wiley & Sons.

Bloomfield, G. [1981] 1988. *Baal Belbora: The End of the Dancing*. Chippendale, New South Wales: Alternative Publishing Co-operative.

Bogen, J. E. 1969. "The Other Side of the Brain II. An Appositional Mind." *Bulletin of the Los Angeles Neurological Societies* 34: 135–62.

———. 1973. "Hemispheric Specificity, Complementarity, and Self-Referential Mappings." *Proceeds of the Society for Neuroscience* 3: 341.

———. 1977. "Some Educational Implications of Hemispheric Specialization." Pp. 133–52 in *The Human Brain*, ed. M. C. Wittrock. Englewood Cliffs, N.J.: Prentice Hall.

Bogen, J. E., R. DeZure, W. D. TenHouten, and J. F. Marsh Jr. 1972. "The Other Side of the Brain IV. The A/P Ratio." *Bulletin of the Los Angeles Neurological Societies* 7: 49–61.

Bouma, A. 1990. *Lateral Asymmetries and Hemispheric Specialization: Theoretical Models and Research*. Amsterdam: Swets and Zeitlinger.

Bourdieu, P. 1977. *Outline of a Theory of Practice*. Translated by R. Nice. New York: Cambridge University Press.

Boyle, R. 1985. "The Dark Side of Mead: Neuropsychological Foundations for Immediate Experience and Mystical Consciousness." Pp. 59–78 in *Studies in Symbolic Interaction*, vol. 6, ed. N. K. Denzin. Greenwich, Conn.: JAI Press.

Bradshaw, J. L., and N. C. Nettleton. 1983. *Human Cerebral Asymmetry*. Englewood Cliffs, N.J.: Prentice Hall.

Brandon, S. G. 1965. *History, Time, and Deity: A Historical and Comparative Study of the Conception of Time in Religious Thought and Practice*. New York: Barnes & Noble.

Broca, P. 1873. "Sur la Mensuration de la Capacité du Crâne." *Memorie Société Anthropologie*, 2d series, vol. 1.

Brown, J. W. 1982. "Hierarchy and Evolution in Neurolinguistics." Pp. 447–67 in *Neural Models of Language Processing*, ed. D. Caplan and J. Marshall. New York: Academic Press.

Bub, D. N., and J. Lewine. 1988. "Different Modes of Word Recognition in the Left and Right Visual Fields." *Brain and Language* 33: 161–88.

Buchanan, C. 1988. "Sexism and Racism." Pp. 259–65 in *The Indigenous Voice: Visions and Realities*, vol. 2, ed. Roger Moody. London: Zed Books.

Bull, W. E. 1968. *Time, Tense, and the Verb: A Study in Theoretical and Applied Linguistics, with Particular Attention to Spanish*. Berkeley: University of California Press.

Butlin, N. G. 1983. *Our Original Aggression: Aboriginal Populations of Southeastern Australia 1788–1850*. Sydney and Boston: G. Allen & Unwin.

Cacioppo, J. T., G. G. Berntson, R. Adolphs et al., eds. 2002. *Foundations in Social Neuroscience*. Cambridge: MIT Press.

Carmon, A., and I. Nachson. 1971. "Effect of Unilateral Brain Damage on Perception of Temporal Order." *Cortex* 7: 410–18.

Cartmill, G. 1955. "Aboriginal Fights and Corroborees." *Grafton Examiner* (August 5): 7.

Chance, M., G. Emory, and R. Payne. 1977. "Status Referents in Long-Tailed Macaques (*Macaca fascicularis*); Precursors and Effects of a Female Rebellion." *Primates* 18: 611–32.

Chance, M. R. A., ed. 1988. *Social Fabrics of the Mind*. Hillsdale, N.J.: Lawrence Erlbaum Associates.

Chance, M. R. A., and C. Jolly. 1970. *Social Groups of Monkeys, Apes, and Men*. New York: E. P. Dutton.

Chase, A. K., and J. R. von Sturmer. 1973. "'Mental Man' and Social Evolutionary Theory." Pp. 3–15 in *The Psychology of Aboriginal Australians*, ed. G. E. Kearney, P. R. de Lacey, and G. R. Davidson. Sydney: John Wiley and Sons Australasia.

Chen, R. S. 1992. *A Comparative Study of Chinese and Western Cyclical Myths*. New York: Peter Lang.

Cheney, D. L., and R. M. Seyfarth. 1990. *How Monkeys See the World*. Chicago: University of Chicago Press.

Clarke, S., G. Assal, and N. de Tribolet. 1993. "Left Hemisphere Strategies in Visual Recognition, Topographic Orientation, and Time Planning." *Neuropsychologia* 31: 99–113.

Cohen, G. 1973. "Hemispheric Differences in Serial Versus Parallel Processing." *Journal of Experimental Psychology* 97: 349–56.

Collier, J. 1911. *The Pastoral Age in Australasia*. London: Whitcombe and Tombs.

Collins, R. 1990. "Stratification, Emotional Energy, and the Transient Emotions." Pp. 27–57 in *Research Agendas in the Sociology of Emotions*, ed. T. D. Kemper. Albany: State University of New York Press.

Corballis, M. C. 1991. *The Lopsided Ape: Evolution of the Generative Mind*. Oxford: Oxford University Press.

Corballis, M., and I. L. Beale. 1971. "On Telling Left from Right." *Scientific American* 224(3): 96–104.

Coren, S. 1992. *The Left-Hander Syndrome: The Causes and Consequences of Left-Handedness*. New York: Macmillan.

Corwin, R. G. 1965. *A Sociology of Education: Emerging Pattern of Class, Status, and Power in the Public Schools*. New York: Appleton-Century-Crofts.

Cowan, J. 1992. *Mysteries of the Dream-Time: The Spiritual Life of Australian Aborigines*. Rev. ed. Lindfield, New South Wales: Unity Press.

Dark, K. R. 1998. *The Waves of Time: Long-Term Change and International Relations*. New York: Pinter.

Das, J. P. 1984. "Aspects of Planning." Pp. 35–50 in *Cognitive Strategies and Educational Performance*, ed. J. R. Kirby. Orlando, Fla.: Academic Press.

Davidson, G. R. 1979. "An Ethnographic Psychology of Aboriginal Cognitive Abilities." *Oceania* XLIX: 270–94.

Davidson, G. R., and L. Z. Klich. 1980. "Cultural Factors in the Development of Temporal and Spatial Ordering." *Child Development* 51: 569–71.

Davis, S. 1989. *Man of All Seasons*. North Ryde, New South Wales, Australia: Angus & Robertson.

Dehaene, S. 1997. *The Number Sense: How the Mind Creates Mathematics*. New York: Oxford University Press.

Dehaene, S., G. Dehaene-Lambertz, and L. Cohen. 1998. "Abstract Representation of Numbers in the Animal and the Human Brain." *Trends in Neuroscience* 21: 355–61.

Delamont, S., and M. Galton. 1986. *Inside the Secondary Classroom*. London: Routledge & Kegan Paul.

Dember, W. M., and J. S. Warm. [1960] 1979. *Psychology of Perception*. 2d ed. New York: Holt, Rinehart, and Winston.

De Renzi, E., and H. Spinnler. 1966. "Visual Recognition in Patients with Unilateral Cerebral Disease." *Journal of Nervous and Mental Disease* 142: 515–25.

de Sousa, R. 1987. *The Rationality of Emotion*. Cambridge: MIT Press.

Dewey, J. 1925. *Experience and Nature*. Chicago and London: Open Court.

———. 1929. *The Quest for Certainty*. New York: Minton, Balch & Company.

———. [1934] 1958. *Art As Experience*. New York: G. P. Putnam's Sons, Capricorn Books.

Dixon, R. M. W. 1966. "Mbabaram: A Dying Australian Language." *Bulletin of the School of Oriental and African Studies* 29: 97–121.

Dōgen Kigen. 1972. *Shōbōgenzō*. 2 vols. Edited by T. Tōru and M. Yaoko. Translated by S. Heine. Tokyo: Inwanami Shoten.

Doob, L. W. 1960. *Becoming More Civilized: A Psychological Exploration*. New Haven, Conn.: Yale University Press.

Doolan, J. K. 1979. "Aboriginal Concept of Boundary: How Do Aborigines Conceive 'Easements'—How Do They Grant Them?" *Oceania* XLIX: 161–68.

Douglas, M. 1975. *Implicit Meanings*. London: Routledge & Kegan Paul.

Dubinskas, F. A., and S. Traweek. 1984. "Closer to the Ground: A Reinterpretation of Walbiri Iconography." *Man*, n.s., 19: 15–30.

Duncan, A. 1980. "Socialization of the Aboriginal Child." *Leader* 2: 50–58.

Duncan, P. 1986. *Shame in Australia*. Unpublished bachelor of letters thesis, Department of Prehistory and Anthropology. Canberra: Australian National University.

Durkheim, É. [1893] 1933. *The Division of Labor in Society*. Translated by G. Simpson. Glencoe, Ill.: Free Press of Glencoe.

———. [1912] 1965. *The Elementary Forms of the Religious Life*. Translated by J. W. Swain. New York: The Free Press.

Durkheim, É., and M. Mauss [1903] 1963. *Primitive Classification*. Translated, edited, and Introduction by R. Needham. Chicago: University of Chicago Press.

Edwards, A. L. 1957. *Techniques of Attitude Scale Construction*. New York: Appleton-Century-Crofts.

Edwards, C., and P. Read. 1989. *The Lost Children: Thirteen Australians Taken from Their Aboriginal Families Tell of the Struggle to Find Their Natural Parents*. Sydney: Doubleday.

Efron, R. 1963. "The Effect of Handedness on the Perception of Simultaneity and Temporal Order." *Brain* 86: 261–84.

———. 1990. *The Decline and Fall of Hemispheric Specialization*. Hillsdale, N.J.: Lawrence Erlbaum.

Ehrenwald, J. 1984. *Anatomy of Genius: Split Brains and Global Minds*. New York: Human Sciences Press.

Eisendrath, C. R. 1971. *The Unifying Moment: The Psychological Philosophy of William James and Alfred North Whitehead*. Cambridge: Harvard University Press.

Elder, B. 1988. *Blood on the Wattle: Massacres and Maltreatment of Australian Aborigines Since 1788*. French Forest, New South Wales: Child & Associates.

Eliade, M. 1951. "Einführende Betrachtung über den Schamanismus." *Paudeuma* 5: 87–97.

———. [1954] 1991. *The Myth of the Eternal Return, or, Cosmos and History*. Translated by W. R. Trask. Bollinger Series XLVI. Princeton: Princeton University Press.

———. 1959. *The Sacred and the Profane: The Nature of Religion*. Translated by W. R. Trask. San Diego: Harcourt Brace Jovanovich.

———. 1963. *Myth and Reality*. New York: Harper & Row.

———. 1973. *Australian Religions: An Introduction*. Ithica and London: Cornell University Press.

Elias, N. 1993. *Time: An Essay*. Translated by E. Jephcott. Oxford: Blackwell.

Elkin, A. P. 1935. "Civilized Aborigines and Native Culture." *Oceania* 6: 129–32.

———. [1938] 1979. *The Australian Aborigines*. Fully rev. ed. London: Angus & Robertson.

———. [1945] 1977. *Aboriginal Men of High Degree*. Sydney: Australasian Publishing.

———. 1954. *The Australian Aborigines and How to Understand Them*. Sydney: Angus and Robertson.

———. 1969. "Elements of Australian Aboriginal Philosophy." *Oceania* 40: 85–98.

Ellis, A. W., A. W. Young, and C. Anderson. 1988. "Modes of Word Recognition in the Left and Right Cerebral Hemispheres." *Brain and Language* 35: 254–73.

Ellis, D. P. 1971. "The Hobbesian Problem of Order: A Critical Appraisal of the Normative Solution." *American Sociological Review* 36: 692–703.

Emory, G. R. 1988. "Social Geometry and Cohesion in Three Primate Species." Pp. 47–60 in *Social Fabrics of the Mind*, ed. M. R. A. Chance. Hillsdale, N.J.: Lawrence Erlbaum.

Evans-Pritchard, E. E. 1940. *The Neur*. Oxford: Clarendon Press.

Fabbro, F. 1994. "Left and Right in the Bible from a Neuropsychological Perspective." *Brain and Cognition* 24: 161–83.

Ferrarotti, F. 1990. *Time, Memory, and Society*. New York: Greenwood Press.

Fiske, A. P. 1991. *Structures of Social Life: The Four Elementary Forms of Human Relations*. New York: The Free Press.

———. 1992. "The Four Elementary Forms of Sociality: Framework for a Unified Theory of Social Relations." *Psychological Review* 99: 689–723.

Fiske, J. 1893. *Man's Destiny*. London: Macmillan.

Flaherty, M. G. 1991. "The Perception of Time and Situated Engrossment." *Social Psychology Quarterly* 54: 76–85.

———. 1999. *A Watched Pot: How We Experience Time*. New York: New York University Press.

Fodor, J. 1983. *The Modularity of Mind*. Cambridge: MIT Press.

Foucault, M. 1977. *Discipline and Punish: The Birth of the Prison*. Translated by A. Sheridan. New York: Pantheon.

Frank, L. K. 1939. "Time Perspectives." *Journal of Social Philosophy* 4: 293–312.

Franz, M.-L. von. 1974. *Number and Time: Reflections Leading toward a Unification of Depth Psychology and Physics*. Evanston, Ill.: Northwestern University Press.

Frazer, J. G. 1961. "Types of Magic." Pp. 1077–87 in *Theories of Society*, vol. 2, ed. T. Parsons, E. Shils, K. D. Naegele, and J. R. Pitts. New York: The Free Press.

Freeman, N. H. 1975. "Temporal and Spatial Ordering in Recall by Five- to Eight-Year-Old Children." *Child Development* 46: 237–39.

Friedland, R., and D. Boden, eds. 1994. *Now Here: Space, Time, and Modernity*. Berkeley: University of California Press.

Fuster, J. M. 1980. *The Prefrontal Cortex: Anatomy, Physiology, and Neuropsychology of the Frontal Lobes*. New York: Raven.

Gale, I. F. 1980. "Aboriginal Time: Dawns of Many Winters." *Folklore* 91: 3–10.

Gallacher, J. D. 1969. "Some Problems Facing an Educator in a Programme of Social Change." Pp. 99–104 in *Aborigines and Education*, ed. S. S. Dunn and C. M. Tatz. Melbourne: Sun Books.

Galton, F. 1880. "Visualized Numerals." *Nature* 21: 536: 323.

Gell, A. 1996. *The Anthropology of Time: Cultural Constructions of Temporal Maps and Images*. Washington, D.C.: Berg.

Gellner, E. 1964. *Thought and Change*. Chicago: University of Chicago Press.

Geschwind, N., and A. M. Galaburda. 1987. *Cerebral Lateralization: Biological Mechanisms, Associations, and Pathology*. Cambridge: MIT Press.

Giddens, A. 1981. *A Contemporary Critique of Historical Materialism*. Vol. 1: *Power, Property, and the State*. London: Macmillan.

Gilbert, P. 1984. *Depression: From Psychology to Brain State*. London: Lawrence Erlbaum.

Gillen, F. J. 1896. "Note on Some Manners and Customs of the Aborigines of the McDonnell Ranges Belonging to the Arunta Tribe." Pp. 159–96 in *Report of the Work of the Horn Scientific Expedition to Central Australia*, 4 vols., ed. B. Spencer. London: Dulau and Company.

Goffman, I. 1961. *Asylums*. Garden City, N.Y.: Anchor Books.

———. 1967. *Interaction Ritual: Essays on Face-to-Face Interaction*. Chicago: University of Chicago Press.

Gondarra, D. 1988. "Father You Gave Us the Dreaming." *Compass Theology Review* 22: 6–8.

Goody, J. 1977. *The Domestication of the Savage Mind*. Cambridge: Cambridge University Press.

Gosden, C. 1994. *Time and Social Being*. Cambridge: Blackwell.

Gould, S. J. 1981. The *Mismeasure of Man*. New York and London: W. W. Norton & Company.

Gouldner, A. W. 1960. "The Norm of Reciprocity: A Preliminary Statement." *American Sociological Review* 25: 161–78.

Grayden, W. 1957. *Adam and Atoms*. Perth: Daniels.

Greenfield, P. 1991. "Language, Tool, and Brain: The Ontogeny and Phylogeny of Hierarchically Organized Sequential Behavior." *Behavioral and Brain Sciences* 14: 531–51.

Grey, A. 1975. *Aboriginal Family Education Centre (A.F.E.C.): A Final Report to the Bernard Van Leer Foundation 1969–1973.* Sydney: Department of Adult Education, University of Sydney.

Gudschinsky, S. C. 1956. "The ABC's of Lexicostatistics (Glottochronology)." *Word* 12: 175–210.

Hall, E. T. [1959] 1973. *The Silent Language.* Garden City, N.Y.: Anchor Press/Doubleday.

Halpern, F. 1973. *Survival: Black/White.* New York: Pergamon Press.

Halstead, W. C. 1947. *Brain and Intelligence: A Quantitative Study of the Frontal Lobes.* Chicago: University of Chicago Press.

Hardy, Sir A. 1965. *The Living Stream: A Restatement of Evolution Theory and Its Relations to the Spirit of Man.* London: Collins.

Harris, J. W. 1979. "Ethnoscience and Its Relevance for Education in Traditional Aboriginal Communities." Master's thesis, University of Queensland.

———. 1990. *One Blood: 200 Years of Aboriginal Encounter with Christianity: A Story of Hope.* Sutherland, New South Wales: Albatross Books.

Harris, P. 1984. "Teaching about Time in Tribal Aboriginal Communities." *Mathematics in Aboriginal Schools Project.* 2. Darwin: Northern Territory Department of Education.

Harris, Stephen 1984. *Culture and Learning: Tradition and Education in North-East Arnhem Land.* Canberra, Australian Capital Territory: Australian Institute of Aboriginal Studies.

Harris, Stewart. 1972. *This Our Land.* Canberra, Australian Capital Territory: Australian National University Press.

Hassard, J., ed. 1990. *The Sociology of Time.* London: Macmillan.

Hausfeld, R. G. 1967. "Hypothetical Futures and Goal Achievement." Paper presented at Australian School of Pacific Administration, Sydney.

Heidegger, M. [1927] 1962. *Being and Time.* Translated by J. Macquarrie and E. Robinson. Oxford: Blackwell.

Heine, S. 1985. *Existential and Ontological Dimensions of Time in Heidegger and Dōgen.* Albany: State University of New York Press.

Hellige, J. B. 1993. *Hemispheric Asymmetry: What's Left and What's Right.* Cambridge: Harvard University Press.

Hertz, R. [1909] 1960. *Death and the Right Hand.* Translated by R. Needham. Aberdeen: Cohen & West.

Hodgson, D. 1991. *The Mind Matters.* Oxford: Oxford University Press.

———. 1996. "Nonlocality, Local Indeterminism, and Consciousness." *Ratio* 9: 1–22.

Holm, N. 1974. "Future Time Perspective in Children from Three Northern Territory Aboriginal Communities." *Developing Education* (February): 1–6.

Howitt, A. W. 1825. "Howitt Papers." Box 3, folder 7, paper 8. Edited by J. F. Uniacke. Canberra, Australian Capital Territory: Australian Institute of Aboriginal Studies.

———. 1904. *The Native Tribes of South-East Australia*. London: Macmillan.

Hubert, H. 1905. "Etude Sommaire de la Représentation du Temps dans la Religion et la Magie." *Annuaire de l'Ecole Partique des Hautes Studies* 1–39.

Hubert, H., and M. Mauss. 1909. *Mélanges d'Historire des Religions*. Paris: Alcan.

Hughes, R. 1987. *The Fatal Shore*. New York: Alfred A. Knopf.

Husserl, E. [1924] 1960. *Cartesian Meditations: An Introduction to Phenomenology*. Translated by D. Cairns. The Hague: Martinus Njjhoff.

Huxley, A. 1946. *The Perennial Philosophy*. London: Chatto & Windus.

Huxley, T. H. [1863] 1906. *Man's Place in Nature*. London: Dent.

Isambert, F.-A. 1979. "*Henri Hubert et la Sociologie du Temps.*" *Revue Française de Sociologie* 20: 183–204.

Isbell, L. A. 1990. "Sudden Short-Term Increase in Mortality of Vervet Monkeys (*Cercopithecus aethiops*) Due to Leopard Predation in Amboseli National Park, Kenya." *American Journal of Primatology* 21: 41–52.

Itani, J. 1984. "Inequality versus Equality for Coexistence in Primate Societies." Pp. 16–189 in *Absolute Values and the New Cultural Revolution: Commemorative Volume of the Twelfth International Conference on the Unity of the Sciences*, ed. Chicago: I.C.U.S. Books.

———. 1988. "The Origin of Human Equality." Pp. 13–156 in *Social Fabrics of the Mind*, ed. M. R. A. Chance. Hillsdale, N. J.: Lawrence Erlbaum.

Jacobs, J. M. 1988. "The Construction of Identity." Pp. 31–43 in *Past and Present: The Construction of Aboriginality*, ed. J. R. Beckett. Canberra, Australian Capital Territory: Aboriginal Studies Press.

James, W. 1890. *The Principles of Psychology*. 2 vols. New York: Henry Holt and Company.

Jaynes, J. 1977. *The Origin of Consciousness in the Breakdown of the Bicameral Mind*. Boston: Houghton Mifflin.

Jung, C. G. 1923. *Psychological Types*. London: Pantheon Books.

Kaberry, P. 1939. *Aboriginal Women, Sacred and Profane*. London: Routledge.

Kalweit, H. 1988. *Dreamtime and Inner Space: The World of the Shaman*. Boston and London: Shambhala.

Kaplan, C. D. 1982. "Addict Life Stories: An Exploration of the Methodological Grounds for the Study of Social Problems." *International Journal of Oral History* 3: 31–50.

Kearney, G., and D. Fitzpatrick. 1976. "Five Psychological Variables Which Distinguish between Two Groups of Aboriginal Australians with Different Levels of Acculturation." Pp. 357–74 in *Aboriginal Cognition: Retrospect and Prospect*, ed. G. E. Kearney and D. W. McElwain. Canberra Australian Capital Territory: Australian Institute of Aboriginal Studies.

Kearney, M. 1984. *World View*. Novato, Calif.: Chandler & Sharp.

Keen, I. 1994. *Knowledge and Secrecy in an Aboriginal Religion*. Oxford: Clarendon Press.

Kern, S. 1983. *The Culture of Time and Space 1880–1918*. Cambridge: Harvard University Press.

Kim, H.-j. 1975. *Dōgen Kigen—Mystical Realist*. Tucson: University of Arizona Press.

Kirk, R. L. 1981. *Aboriginal Man Adapting: The Human Biology of Australian Aborigines*. Oxford: Oxford University Press.

Klatzky, R. L., and R. C. Atkinson. 1971. "Specialization of the Cerebral Hemispheres in Scanning for Information in Short Term Memory." *Perception and Psychophysics* 10: 335–38.

Kluckhohn, F. R., and F. L. Strodbeck. 1961. *Variation in Value Orientations*. Evanston, Ill.: Row-Peterson.

Kohli, M. 1986. "Biographical Research in the German Language Area." Pp. 91–110 in *Commemorative Book in Honor of Florian Znaniecki on the Centenary of His Birth*, ed. Z. Dulczewsk. Poznan, Poland: Natukowe.

Kolig, E. 1981. *The Silent Revolution: The Effects of Modernization on Australian Aboriginal Religion*. Philadelphia: Institute for the Study of Human Issues.

Kunzig, R. 1997. "A Head for Numbers." *Discovery* 18 (July): 108–15.

Lakoff, G., and M. Johnson. 1980. *Metaphors We Live By*. Chicago: University of Chicago Press.

Langer, E. 1997. *The Power of Mindful Learning*. Reading, Mass.: Addison-Wesley.

Langer, W. L. 1935. *The Diplomacy of Imperialism, 1890–1902*. New York and London: Alfred A. Knopf.

Larson, D. W. 1988. "The Psychology of Reciprocity in International Relations." *Negotiations Journal* 4: 281–301.

Laughlin, C. D. 1988. "The Prefrontosensorial Polarity Principle: Toward a Neurophenomenological Theory of Intentionality." *Biology Forum* 81: 245–62.

LeDoux, J. 1996. *The Emotional Brain: The Mysterious Underpinnings of Emotional Life*. New York: Simon & Schuster.

Lee, D. 1949. "Being and Values in a Primitive Culture." *The Journal of Philosophy* 46: 401–15.

Leeuw, G. van der. 1957. "Primordial Time and Final Time." In *Man and Time: Papers from the Eranos Yearbooks*, vol. 3, ed. H. Corbin et al. New York: Pantheon Books.

Lestienne, R. 1995. *The Children of Time: Causality, Entropy, Becoming*. Translated by E. C. Neher. Urbana and Chicago: University of Illinois Press.

Lévi-Strauss, C. 1965. "The Structural Study of Myth." Pp. 81–106 in *Myth: A Symposium*, ed. T. A. Sebeok. Bloomington: Indiana University Press.

———. 1966. *The Savage Mind*. Chicago: University of Chicago Press.

Levy-Agresti, J., and R. Sperry. 1968. "Differential Perceptual Capacities in Major and Minor Hemispheres." *Proceedings of the National Academy of Science* 61: 1151.

Lévy-Bruhl, L. [1910] 1985. *How Natives Think*. Translated by L. A. Clark. Princeton: Princeton University Press.

Lewis, D. 1976. "Observations of Route Finding and Spatial Orientation among the Aboriginal People of the Western Desert Region of Central Australia." *Oceania* XLVI: 249–82.

Lewis, W. 1927. *Time and Western Man*. London: Chatto and Windus.

Liberman, K. 1985. *Understanding Interaction in Central Australia: An Ethnomethodological Study of Australian Aboriginal People*. Boston: Routledge & Kegan Paul.

Lovejoy, F. 1975. "The Relevance of Aboriginal Concepts of Time to Assimilation Policy and Practice." Paper presented at the 46th Congress of the Australian and New Zealand Association for the Advancement of Science, January, Canberra, Australian Capital Territory.

Luhmann, N. 1982. *The Differentiation of Society*. Translated by S. Holmes and C. Larmore. New York: Columbia University Press.

Luria, A. R. 1966. *Higher Cortical Functions in Man*. New York: Basic Books.

———. 1969. "Frontal Lobe Syndromes." Pp. 725–57 in *Handbook of Clinical Neurology*, vol. 2, ed. P. J. Vinkin and G. W. Bruyn. New York: John Wiley & Sons.

———. 1973. *The Working Brain: An Introduction to Neuropsychology*. Translated by B. Haigh. London: Allen Lane, Penguin Press.

———. 1982. *Language and Cognition*. New York: Wiley.

MacLean, P. D. 1964. "Mirror Display in the Squirrel Monkey (*Siamiri Sciureus*)" *Science* 146: 950–52.

———. 1973. *A Triune Concept of the Brain and Behavior*. Toronto: University of Toronto Press.

———. 1977. "The Triune Brain in Conflict." *Psychotherapy and Psychosomatics* 28: 207–20.

Maddock, K. 1972. *The Australian Aborigines: A Portrait of Their Society*. London: Allen Lane, Penguin Press.

Magoun, H. W. 1963. *The Waking Brain*. 2d ed. Springfield, Ill.: Thomas.

Maines, D. R., N. M. Sugrue, and M. A. Katovich. 1983. "The Sociological Import of G. H. Mead's Theory of the Past." *American Sociological Review* 48: 161–73.

Malinowski, B. [1922] 1961. *Argonauts of the Western Pacific: An Account of Native Enterprise and Adventure in the Archipelagoes of Melanesian New Guinea*. New York: E. P. Dutton & Co.

———. 1926. *Crime and Custom in Savage Society*. New York: Harcourt, Brace & Company.

Mann, H., M. Siegler, and H. Osmond. 1968. "The Many Worlds of Time." *Journal of Analytic Psychology* 13: 33–56.

Marshack, A. 1972. *The Roots of Civilization*. New York: McGraw-Hill.

Marx, K. 1887. *Capital*. Vols. 1–3. Translated by S. Moore and E. Aveling. Moscow: Progress Publishers.

Masuzawa, T. 1993. *In Search of Dreamtime: The Quest for the Origin of Religion*. Chicago: University of Chicago Press.

Mattingley, C., ed., and K. Hampton, co-ed. 1988. *Survival in Our Own Land: "Aboriginal" Experience in "South Australia" since 1836*. Sydney: Hodder and Stoughton.

Mauss, M. [1925] 1954. *The Gift: Forms and Functions of Exchange in Archaic Societies*. Translated by I. Cunnison. London: Cohen & West.

McConvell, P. 1985. "Time Perspective in Aboriginal Australian Culture: Two Approaches to the Origin of Subsections." *Aboriginal History* 9: 53–79.

McElwain, N. 1988. "Seneca Iroquois Concepts of Time." *Cosmos 4 Amerindian Cosmology*. Edinburgh: Traditional Cosmological Society.

Mead, G. H. 1929. "The Nature of the Past." Pp. 235–42 in *Essays in Honor of John Dewey*, ed. J. Cross. New York: Henry Holt.

———. [1932] 1980. *The Philosophy of the Present*. Edited by A. E. Murray. Prefatory remarks by J. Dewey. Chicago and London: Open Court.

———. 1934. *Mind, Self, and Society*. Chicago: University of Chicago Press.

Meehan, B. 1982. *Shell Bed to Shell Midden*. Canberra, Australian Capital Territory: Australian Institute of Aboriginal Studies.

Meggitt, M. J. 1962. *Desert People: A Study of the Walbiri Aborigines of Central Australia*. Sydney: Angus & Robertson.

Merleau-Ponty, M. 1962. *Phenomenology of Perception*. Translated by C. Smith. London: Routledge & Kegan Paul.

Milner, B. 1964. "Some Effects of Frontal Lobectomy in Man." Pp. 313–34 in *The Frontal Granular Cortex and Behavior: A Symposium*, ed. J. M. Warren and K. Akert. New York: McGraw-Hill.

Minkowski, E. [1933] 1970. *Lived Time: Phenomenological and Psychopathological Studies*. Evanston, Ill.: Northwestern University Press.

Mol, H. 1982. *The Firm and the Formless: Religion and Identity in Aboriginal Australia*. Waterloo, Ontario, Canada: Wilfred Laurier University Press.

Montagner, H., A. Restoin, D. Rodriguez, V. Ullman, M. Viala, D. Laurent, and D. Goddard. 1988. "Social Interaction of Young Children with Peers and Their Modification in Relation to Environmental Factors." Pp. 237–59 in *Social Fabrics of the Mind*, ed. M. R. A. Chance. Hillsdale, N.J.: Lawrence Erlbaum.

Morphy, H. 1984. *Journey to the Crocodile's Nest: An Accompanying Monograph to the Film Madaarpa at Gurka'wuy*. Canberra, Australian Capital Territory: Australian Institute of Aboriginal Studies.

———. 1990. "Myth, Totemism, and the Creation of Clans." *Oceania* 60: 312–29.

Morris, W., ed. 1970. *The American Heritage Dictionary of the English Language.* Boston: Houghton Mifflin.

Morton, J. 1987. "Singing Subjects and Sacred Objects: More on Munn's 'Transformations of Subjects into Objects' in Central Australian Myth." *Oceania* 58: 100–18.

Moyer, R., and T. K. Landauer. 1967. "Time Required for Judgment of Numerical Inequality." *Nature* 215:5109: 1519–20.

Mumford, L. [1934] 1955. "The Monastery and the Clock." Pp. 3–10 in *The Human Prospect*, ed. H. T. Moore and K. W. Deutsch. Boston: Beacon Press.

Munn, N. 1969. "The Effectiveness of Symbols in Murngun Rite and Myth." Pp. 178–207 in *Forms of Symbolic Action: Proceedings of the 1969 Spring Meeting of the American Ethnological Society*, ed. R. F. Spencer. Seattle: University of Washington Press.

———. 1970. "The Transformation of Subjects into Objects in Walbiri and Pitjantjatjara Myth." Pp. 141–63 in *Australian Aboriginal Anthropology: Modern Studies in the Social Anthropology of the Australian Aborigines*, ed. R. M. Berndt. Nedlands: University of Western Australia Press.

———. [1973] 1986. *Walbiri Iconography: Graphic Representation and Cultural Symbolism in a Central Australian Society.* Ithica and London: Cornell University Press.

Myers, F. 1972. "What Men Do: A Preliminary Consideration of the Classification of the Self in the Universe in Aboriginal Culture." Master's thesis, Bryn Mawr College, Pennsylvania.

———. 1979. "Emotions and the Self: A Theory of Personhood and Political Order among Pintupi Aborigines." *Ethos* 7: 343–70.

———. 1986. *Pintupi Country, Pintupi Self: Sentiment, Place, and Politics among Western Desert Aborigines.* Canberra, Australian Capital Territory: Australian Institute of Aboriginal Studies.

———. 1988. "Burning the Truck and Holding the Country: Property, Time, and the Negotiation of Identity among Pintupi Aborigines." Pp. 52–74 in *Hunters and Gatherers 2: Property, Power, and Ideology*, ed. T. Ingold, D. Riches, and J. Woodburn. New York: St. Martin's Press.

Nass, R., and M. S. Gazzaniga. 1987. "Cerebral Lateralization and Lateralization in Human Central Nervous Systems." Pp. 701–61 in *Handbook of Physiology*, vol. 5, part 2, *The Nervous System*, ed. V. B. Mountcastle and S. R. Geiger. Baltimore: Waverly Press.

Newcombe, F. 1969. *Missile Wounds of the Brain: A Study of Psychological Deficits.* London: Oxford University Press.

Newton, I. [1729] 1934. *The Mathematical Principles of Natural Philosophy and His System of the World.* Translated by A. Motte. Berkeley: University of California Press.

Noble, J. 1968. "Paradoxical Interocular Transfer of Mirror-Image Discriminations in the Optic Chiasm Sectioned Monkey." *Brain Research* 10: 127–51.

Nowotny, H. 1975. "Time Structuring and Time Measurement: On the Interrelations between Timekeeping and Social Time." Pp. 325–61 in *The Study of Time*, vol. 2, ed. J. T. Fraser and N. Lawrence. New York: Springer Verlag.

———. 1985. "From the Future to the Extended Present-Time in Social Systems." Pp. 1–21 in *Time Preference: An Interdisciplinary Theoretical and Empirical Approach*, ed. G. Kirsch, P. Nijkamp, and K. Zimmerman. Berlin: Wissenschaftszentrum.

Oates, L. F. 1975. *The 1973 Supplement to a Revised Linguistic Survey of Australia*. 2 vols. Armidale, New South Wales: Armidale Christian Book Center.

O'Connor, N., and B. M. Hermelin. 1973a. "Short-Term Memory for the Order of Pictures and Syllables by Deaf and Hearing Children." *Neuropsychologia* 11: 437–42.

———. 1973b. "The Spatial or Temporal Organization of Short-Term Memory." *Quarterly Journal of Experimental Psychology* 25: 335–43.

O'Malley, P. 1989. "Gentle Genocide: The Government of Aboriginal People in Central Australia." *Social Justice* 21: 46–65.

Orton, S. B. 1937. *Reading, Writing, and Speech Problems in Children*. New York: Norton.

Ortony, A., and T. J. Turner. 1990. "What's Basic about Basic Emotions?" *Psychological Review* 9: 315–31.

Oxford English Dictionary, the Compact Edition. 1971. vols. 1–2. Melbourne: Oxford University Press.

Passingham, R. E. 1973. "Anatomical Differences between the Neocortex of Man and Other Primates." *Brain, Behavior, and Evolution* 7: 337–59.

Phillip, A. 1914. "Dispatch No. 1 to Lord Sydney, 15th May 1788." *Historical Records of Australia* (ser. 1): 1: 16–34.

Piaget, J. [1952] 1963. *The Origins of Intelligence in Children*. New York: W. W. Norton & Company.

Plateau, J. A. F. 1872. "*Sur la Measure des Sensations Physiques, et sur la Loi qui lie l'Intensité de ces Sensation à l'Intensité de la Cause Excitante.*" *Bulletin de Academie Royale de Belgique* 33: 376–88.

Plutchik, R. [1962] 1991. *The Emotions: Facts, Theories, and a New Model*. Lanham, Md.: University Press of America.

Polich, J. M. 1980. "Left Hemisphere Superiority for Visual Search." *Cortex* 16: 39–50.

———. 1984. "Hemispheric Patterns in Visual Search." *Brain and Cognition* 3: 128–39.

Postone, M. 1993. *Time, Labor, and Social Domination: A Reinterpretation of Marx's Critical Theory*. New York: Cambridge University Press.

Power, M. 1986. "The Foraging Adaptation of Chimpanzees and the Recent Behavior of the Provisioned Apes in Gombe and Mahale National Parks, Tanzania." *Human Evolution* 1: 251–66.

Pribram, K. H. 1971. *Languages of the Brain: Experimental Paradoxes and Principles in Neuropsychology*. Englewood Cliffs, N.J.: Prentice Hall.

———. 1981. "Emotions." Pp. 102–34 in *Handbook of Clinical Neuropsychology*, vol. 1, ed. S. B. Filskov and T. J. Boll. New York: John Wiley & Sons.

Pribram, K. H., and A. R. Luria, eds. 1973. *Psychophysiology of the Frontal Lobes*. New York: Academic Press.

Pribram, K. H., and W. E. Tubbs. 1967. "Short-Term Memory, Parsing, and the Primate Frontal Cortex." *Science* 156 (3783): 1765–67.

Prigogine, I., and I. Stengers. 1984. *Order out of Chaos: Man's New Dialogue with Nature*. New York: Bantom Books.

Rabbit, P. 1997. "Introduction: Methodologies and Models in the Study of Executive Function." Pp. 1–38 in *Methodology of Frontal and Executive Function*, ed. P. Rabbit. Perth, Western Australia, and East Sussex, United Kingdom: Psychology Press.

Radcliffe-Brown, A. R. 1930. "The Social Organization of Australian Tribes." *Oceania* 1: 34–63.

———. 1945. "Religion and Society." Pp. 153–77 in *Structure and Function in Primitive Society*, ed. A. R. Radcliffe-Brown. London: Cohen & West.

———. 1952. *Structure and Function in Primitive Society*. London: Cohen and West.

Rammsayer, T. H., and W. H. Vogel. 1992. "Pharmacologic Properties of the Internal Clock Underlying Time Perception in Humans." *Neuropsychobiology* 26: 71–80.

Rappaport, R. 1992. "Ritual, Time, and Eternity." *Zygon* 27: 5–30.

Reason, D. 1979. "Classification, Time, and the Organization of Production." Pp. 221–47 in *Classifications in Their Social Context*, ed. R. F. Ellen and D. Reason. London: Academic Press.

Reynolds, H. 1974. "Progress, Morality, and the Dispossession of the Aborigines." *Meanjin Quarterly* 33: 306–12.

Richardson, J. 1986. *Existential Epistemology: A Heideggerian Critique of the Cartesian Project*. Oxford: Clarendon Press.

Ricoeur, P. 1983, 1985, 1988. *Time and Narrative*. Vols. 1–3. Translated by K. McLaughlin and D. Pellauer (originally *Temps et Récit*, Editions du Seuil, 1983). Chicago: University of Chicago Press.

Risberg, J., J. H. Halsey, E. L. Wills, and E. M. Wilson. 1975. "Hemispheric Specialization in Normal Man Studied by Bilateral Measurement of the Regional Cerebral Blood Flow: A Study with the 133-Xe Inhalation Technique." *Brain* 98: 511–24.

Roberts, R. G., R. Jones, and M. A. Smith. 1990. "Thermoluminescence Dating of a 50,000-Year-Old Human Occupation Site in Northern Australia." *Nature* 345: 153–56.

Roget, P. M. [1852] 1977. *Roget's International Thesaurus*. 4th ed. Rev. by R. Chapman. New York: Harper & Row.

Rosaldo, R. 1980. *Ilongot Headhunting 1883–1974: A Study in Society and History*. Stanford: Stanford University Press.

Rose, D. 1992. *Dingo Makes Us Human: Life and Land in an Aboriginal Australian Culture*. Cambridge: Cambridge University Press.

Rose, F. G. G. 1987. *The Traditional Mode of Production of the Australian Aborigines*. North Ryde, New South Wales: Angus & Robertson.

Rotenberg, R. 1992. *Time and Order in Metropolitan Vienna*. Washington and London: Smithsonian Institution Press.

Rowley, C. D. 1986. *Recovery: The Politics of Aboriginal Reform*. Harmondsworth: Penguin.

Rowse, T. 1994. *After Mabo: Interpreting Indigenous Tradition*. Carlton, Victoria: Melbourne University Press.

Rudder, J. 1983. *Qualitative Thinking: An Examination of the Classificatory System, Evaluative System, and Cognitive Structures of the Yolnu people of Northeast Arnhem Land*. Unpublished, Master's thesis. Canberra, Australian Capital Territory: Australian National University.

———. 1991. "Yolngu Time." Unpublished manuscript. Canberra, Australian Capital Territory: Australian University Press.

———. 1993. "Yolnu Cosmology: An Unchanging Cosmos Incorporating a Rapidly Changing World?" Ph.D. thesis in anthropology, Australian National University.

Rutz, H. Z., ed. 1992. *The Politics of Time*. Washington, D.C.: American Anthropological Association.

Sagen, C. 1977. *The Dragons of Eden: Speculations on the Evolution of Human Intelligence*. New York: Random House.

Sahlins, M. 1981. *Historical Metaphors and Mythical Realities: Structure in the Early History of the Sandwich Islands Kingdom*. Ann Arbor: University of Michigan Press.

Sansom, B. 1980. *The Camp at Wallaby Cross*. Canberra, Australian Capital Territory: Australian Institute of Aboriginal Studies.

———. 1988. "The Past Is a Doctrine of Person." Pp. 147–60 in *Past and Present: The Construction of Aboriginality*, ed. J. R. Beckett. Canberra, Australian Capital Territory: Aboriginal Studies Press.

Scheler, M. 1926. *Die Wissenformen und die Gesellschaft* ("The Forms of Knowledge and Society"). Leipzig: Der Neue Geist Verlag.

Schutz, A. 1967. *Collected Papers*. Vol. 1. *The Problem of Social Reality*. Edited by M. Natanson. The Hague, the Netherlands: Martinus Nijhoff.

Schutz, A., and T. Luckmann. 1973. *The Structures of the Life World*. Translated by R. M. Zaner and H. T. Engelhardt Jr. London: Heinemann.

Schwartz, B. 1979. "Waiting, Exchange, and Power: The Distribution of Time in Social Systems." *American Journal of Sociology* 79: 841–70.

———. 1981. *Vertical Classification: A Study in Structuralism and the Sociology of Knowledge*. Chicago: University of Chicago Press.

Seagrim, G., and R. Lendon. 1980. *Furnishing the Mind: A Comparative Study of Cognitive Development in Central Australian Aborigines*. Sydney: Academic Press.

Senate Select Committee on Aborigines and Torres Strait Islanders. 1976. *Environmental Conditions of Aborigines and Torres Strait Islanders and the Preservation of Their Sacred Sites*. Canberra, Australian Capital Territory: Australian Government Publishing Service.

Shellshear, J. L. 1937. *The Brain of the Aboriginal Australian: A Study in Cerebral Morphology*. Philosophical Transactions of the Royal Society of London, Series B-Biological Sciences. London: Harrison & Sons.

Slife, B. D. 1981. "Psychology's Reliance on Linear Time: A Reformulation." *Journal of Mind and Behavior* 2: 27–46.

Smith, A. [1776] 1979. *An Inquiry in the Nature and Causes of the Wealth of Nations*. Oxford: Oxford University Press.

Smyth, R. B. [1878] 1972. *The Aborigines of Victoria*. 2 vols. Melbourne: John Currey O'Neil.

Sommer, B. A. 1969. *Kungen Phonology: Synchronic and Diachronic*. Canberra, Australian Capital Territory: Australian National University Press.

Sommerlad, E. 1976. *Kormilda, the Way to Tomorrow? A Study in Aboriginal Education*. Canberra, Australian Capital Territory: Australian National University Press.

Sousa, R. de. 1987. *The Rationality of Emotions*. Cambridge: MIT Press.

Spencer, W. B., and F. J. Gillin. 1925. *The Arunta: A Study of a Stone-Age People*. 2 vols. London: Macmillan.

Spengler, O. 1926. *The Decline of the West: Form and Actuality*. Translated by C. F. Atkinson. New York: Alfred A. Knopf.

Springer, S. P., and G. Deutsch. 1981. *Left Brain, Right Brain*. San Francisco: W. H. Freeman and Company.

Stambaugh, J. 1990. *Impermanence Is Buddha-Nature: Dōgen's Understanding of Temporality*. Honolulu: University of Hawaii Press.

Stanner, W. E. H. 1958. "Continuity and Change among the Aborigines." *The Australian Journal of Science* 21: 99–109.

———. 1965. "Religion, Totemism, and Symbolism." Pp. 207–37 in *Aboriginal Man in Australia*, ed. R. M. Berndt and C. H. Berndt. Sydney: Angus & Robertson.

———. 1968. *After the Dreaming: Black and White Australians—An Anthropologist's View*. The Boyer Lecture. Sydney: The Australian Broadcasting Commission.

———. 1979. *White Man Got No Dreaming: Essays 1938–1973*. Canberra, Australian Capital Territory: Australian National University Press.

Stevens, S. S. 1960. "The Psychophysics of Sensory Function." *American Scientist* 48: 226–53.

———. 1961. "The Psychophysics of Sensory Function. Pp. 1–33 in *Sensory Communication*, ed. W. A. Rosenblith. Cambridge: M.I.T. Press.

———. 1975. *Psychophysiology: Introduction to Its Perceptual, Neural, and Social Prospects*. New York: John Wiley & Sons.

Strehlow, C. 1907. *Die Aranda- und Loritja-Stamme in Zentral Australien I: Mythen, Sagen und Marchen des Aranda-Stammes*. Frankfurt am Main: Joseph Baer.

———. 1908. *Die Aranda- und Loritja-Stamme in Zentral Australien II: Mythen, Sagen, und Marchen des Loritja-Stammes, die Totemistischen Vorstellengen und die Tjuringa der Aranda und Loritja*. Frankfurt am Main: Joseph Baer.

———. 1913. *Die Aranda- und Loritja-Stamme in Zentral Australien IV: Das Soziale Leben die Aranda und Loritja*. Frankfurt am Main: Joseph Baer.

Strehlow, T. G. H. 1933. "Ankotarinja, an Aranda Myth." *Oceania* 4: 187–200.

———. 1947. *Aranda Traditions*. Melbourne: Melbourne University Press.

———. 1956. *The Sustaining Ideals of an Australian Aboriginal Society*. Adelaide: Aborigines' Advancement League.

———. 1964. "Personal Monototemism in a Polytotemic Community." Pp. 723–54 in *Festschrift für Ad. E. Jensen*, vol. 2, ed. E. Haberland, M. Schuster, and H. Strabe. Munich: Klaus Renner Verlag.

———. 1970. "Geography and the Totemic Landscape in Central Australia: A Functional Study." Pp. 92–140 in *Australian Aboriginal Anthropology: Modern Studies in the Social Anthropology of the Australian Aborigines*, ed. R. M. Berndt. Nedlands: University of Western Australia Press.

———. 1971. *Songs of Central Australia*. Sydney: Angus & Robertson.

Sturmer, J. von. 1981. "Talking with Aborigines." *Australian Institute of Aboriginal Studies Newsletter*, n.s., 15: 13–30.

Stuss, D. T., and D. F. Benson. 1986. *The Frontal Lobes*. New York: Raven Press.

Swain, T. 1991. "Aboriginal Religions in Time and Space." Pp. 151–65 in *Religion in Australia: Sociological Perspectives*, ed. A. W. Black. Sydney: Allen & Unwin.

———. 1993. *A Place for Strangers: Towards a History of Australian Aboriginal Being*. New York: Cambridge University Press.

Talpin, G. et al., with an introductory chapter by J. D. Woods. 1879. *The Native Tribes of South Australia*. Adelaide, Australia: E. S. Wigg & Son.

Tatz, C. 1982. *Aborigines and Uranium and Other Essays*. Melbourne: Heinemann.

TenHouten, W. D. 1985. "Right Hemisphericity of Australian Aboriginal Children: Effects of Culture, Sex, and Age on Performances of Closure and Similarities Tests." *International Journal of Neuroscience* 28: 125–46.

———. 1986. "Right Hemisphericity of Australian Aboriginal Children II: Conjugate Lateral Eye Movements." *International Journal of Neuroscience* 30: 255–60.

———. 1992a. "Cerebral Lateralization: A Scientific Paradigm in Crisis? Critique of Efron and Corballis." *Journal of Social and Evolutionary Systems* 15: 319–26.

———. 1992b. "Into the Wild Blue Yonder: On the Emergence of the Ethnoneurologies." *Journal of Social and Biological Structures* 14: 381–408.

———. 1996. "Outline of a Socioevolutionary Theory of the Emotions." *International Journal of Sociology and Social Policy* 16: 189–208.

———. 1999a. "Explorations in Neurosociological Theory: From the Spectrum of Affect to Time-Consciousness." Pp. 41–80 in *Social Perspectives on Emotions*, vol. 5, ed. D. D. Franks and T. S. Smith. Greenwich, Conn.: JAI Press.

———. 1999b. "The Four Elementary Forms of Sociality, Their Biological Bases, and Their Implications for Affect and Cognition." Pp. 253–84 in *Advances in Human Ecology*, vol. 8, ed. L. Freese. Greenwich, Conn.: JAI Press.

———. 1999c. "Text and Temporality: Patterned-Cyclical and Ordinary-Linear Forms of Time-Consciousness, Inferred from a Corpus of Australian Aboriginal and Euro-Australian Life-Historical Interviews." *Symbolic Interaction* 22: 121–37.

———. 2004. "Time and Society: A Cross-Cultural Study." *Free Inquiry in Creative Sociology* 32: 19–32.

Terdiman, R. 1993. *Present Past: Modernity and the Memory Crisis*. Ithaca, N.Y.: Cornell University Press.

Thompson, E. P. 1967. "Time, Work-Discipline, and Industrial Capitalism." *Past and Present* 36: 52–97.

Thomson, D. F. 1935. "The Joking Relationship and Organized Obscenity in North Queensland." *American Anthropologist*, n.s., 37: 460–89.

Tonkinson, R. 1978. *The Mardudjara Aborigines: Living the Dream in Australia's Desert*. New York: Holt, Rinehart, and Winston.

Tönnies, F. [1887] 2000. *Community and Civil Society*. Translated by J. Harris and M. Hollis. New York: Cambridge University Press.

Treisman, M., A. Faulkner, P. L. Naish, and D. Brogan. 1990. "The Internal Clock: Evidence for a Temporal Oscillator Underlying Time Perception with Some Estimates of Its Characteristic Frequency." *Perception* 19: 705–43.

Trivers, R. L. 1971. "The Evolution of Reciprocal Altruism." *The Quarterly Review of Biology* 46: 35–57.

Turner, D. H. 1980. *Australian Aboriginal Social Organization*. Canberra, Australian Capital Territory: Australian Institute of Aboriginal Studies.

———. 1985. *Life before Genesis: A Conclusion (An Understanding of Australian Aboriginal Culture)*. New York: Peter Lang.

Turner, F. 1985. *Natural Classicism: Essays on Literature and Science*. New York: Paragon House.

Turner, T. S. 1969. "Oedipus: Time and Structure in Narrative." Pp. 26–68 in *Forms of Symbolic Action: Proceedings of the 1969 Annual Spring Meeting of the American Ethnological Society*, ed. R. F. Spencer. Seattle: University of Washington Press.

Vernon, P. E. 1966. "Educational and Intellectual Development among Canadian Indians and Eskimos." *The Education Review* 18: 79–91.

Wake, C. S. 1872. "The Mental Characteristics of Primitive Man, As Exemplified by the Australian Aborigines." *Journal of the Anthropological Institute* 1: 74–84.

Warden, S. A. 1968. *The Leftouts: Disadvantaged Students in Heterogeneous Schools*. New York: Holt, Rinehart, and Winston.

Warner, W. L. [1937] 1958. *A Black Civilization: A Study of an Australian Tribe*. New York: Harper & Brothers.

Weber, M. [1904–1905] 1989. *The Protestant Ethic and the Spirit of Capitalism*. London: Unwin Hyman.

———. [1922] 1978. *Economy and Society*. Translated by G. Roth and C. Wittich. Berkeley: University of California Press.

Welsby, T. 1917. "Recollections of the Natives of Moreton Bay: Together with Some of Their Names and Customs of Living." *Historical Society of Queensland Journal* 1: 110–29.

Whatmore, G., and D. R. Kohli. 1974. *The Physiopathology and Treatment of Functional Disorders*. New York: Grune & Stratton.

White, J. P., and D. J. Mulvaney. 1987. "How Many People?" Pp. 15–17 in *Australians to 1788*, ed. D. J. Mulvaney and J. P. White. Broadway, New South Wales: Fairfax, Syme & Weldon Associates.

Whitehead, A. N. 1929. *Process and Reality*. Cambridge: Cambridge University Press.

Whitrow, G. J. 1972. *The Nature of Time*. New York: Holt, Rinehart, and Winston.

———. 1989. *Time in History: Views of Time from Prehistory to the Present*. Oxford and New York: Oxford University Press.

Whorf, B. L. 1941. "The Relation of Habitual Thought and Behavior to Language." Pp. 75–93 in *Language, Culture, and Personality: Essays in Memory of Edward Sapir*, ed. Leslie Spier. Menasha, Wis.: Sapir Memorial Publication Fund.

Williams, N. 1986. *The Yolngu and Their Land: A System of Land Tenure and the Fight for Its Recognition*. Canberra, Australian Capital Territory: Australian Institute of Aboriginal Studies.

Williams, R. 1977. *Marxism and Literature*. Oxford: Oxford University Press.

Wood, D. 1989. *The Deconstruction of Time*. Atlantic Highlands, N.J.: Humanities Press.

Woodburn, J. 1980. "Hunters and Gatherers Today and Reconstruction of the Past." Pp. 95–117 in *Soviet and Western Anthropology*, ed. E. Gellner. London: Duckworth.

Woodcock, G. 1944. "The Tyranny of the Clock." *Politics* 1: 265–66.

Wurm, S. A. 1972. *Languages of Australia and Tasmania*. The Hague, the Netherlands: Mouton.

Wynn, K. 1996. "Addition and Subtraction By Human Infants." *Nature* 358 (6389): 749–50; "Erratum" *Nature* 361 (6410): 374.

Yallop, C. 1982. *Australian Aboriginal Languages*. London: André Deutsch.

Young, A. W., and A. W. Ellis. 1985. "Different Methods of Lexical Access for Words Presented in the Left and Right Visual Hemifields." *Brain and Language* 24: 326–58.

Young, M. D. 1988. *The Metronomic Society: Natural Rhythms and Human Timetables*. Cambridge: Harvard University Press.

Zelazo, P. D., U. Müller, D. Frye, and S. Marcovitch. 2003. "The Development of Executive Function in Early Childhood." *Monographs of the Society for Research in Child Development* 68 (3, serial number 274).

Zern, D. 1967. "The Influence of Certain Developmental Factors on Fostering the Ability to Differentiate the Passage of Time." *Journal of Social Psychology* 72: 9–17.

Zerubavel, E. 1981. *Hidden Rhythms: Schedules and Calendars in Social Life*. Berkeley: University of California Press.

———. 1991. *The Fine Line: Making Distinctions in Everyday Life*. New York: The Free Press.

———. 1997. *Social Mindscapes: An Invitation to Cognitive Sociology*. Cambridge: Harvard University Press.

Zimmerman, M. E. 1981. *Eclipse of the Self: The Development of Heidegger's Concept of Authenticity*. Athens and London: Ohio University Press.

Znaniecki, F. 1934. *The Method of Sociology*. New York: Farrar & Rinehart.

Name Index

Abram, D., 135, 154–55
Adam, B., 2, 26, 61, 63–64, 171–72, 189, 210
Ahern, C., 180
Altom, M. W., 79
Alverson, H., 192
Anderson, C., 77
Ariotti, P. E., 36
Aristotle, 1, 12, 65, 68, 96
Armstrong, E., 91
Assal, G., 78
Atkinson, R. C., 77
Axelrod, R., 156

Bakan, P., 107
Banks, R., 150
Barbalet, J. M., 92, 157
Barden, G., 48
Barnes, R., 2
Baumgärtner, G., 81
Beale, I. L., 115
Beckett, J. R., 17, 149
Benson, D. F., 91
Bentham, J., 16
Berdyaev, N., 182
Bergland, R., 77
Bergson, H., 59
Berndt, C. H., 122, 144, 146, 158, 177
Berndt, R. M., 122, 144, 146, 158, 177
Bever, T. G., 74
Blau, P. M., 156
Bloomfield, G., 14
Boden, D., 215
Bogen, J. E., 75, 77, 80, 113–14, 215
Bouma, A., 77
Bourdieu, P., 136, 166

Boyle, R., 92–93, 118
Bradshaw, J. L., 74, 76
Brandon, S. G., 165–66
Broca, P., 15
Brown, J. W., 118
Bub, D. N., 77
Buchanen, C., 221n2
Bull, W. E., 37
Butlin, N. G., 18

Cacioppo, J. T., 215
Carmon, A., 78
Cartmill, G., 27
Chance, M. R. A., 4–5, 96–100, 102, 109
Chase, A. K., 12–13
Chen, R. S., 26, 51–52
Cheney, D. L., 98
Clarke, S., 78
Cohen, G., 76
Cohen, L., 78–79
Collier, J., 13
Collins, R., 159
Corballis, M. C., 7, 73–75, 115
Coren, S., 27
Corwin, R. G., 146
Cowan, J., 29, 138, 147–48

Dark, K. R., 215
Das, J. P., 91
Davidson, G. R., 80
Davis, S., 47
De Renzi, E., 124
Dehaene, S., 78–79
Dehaene-Lambertz, G., 79
Delamont, S., 63

Dember, W. M., 70–71
Deutsch, G., 215
Dewey, J., 88, 187
Dixon, R. M. W., 221n3
Dōgen Kigen, 85–86, 135, 145, 151
Doob, L. W., 166
Doolan, J. K., 170, 174
Douglas, M., 36
Dubinkskas, F. A., 23, 32, 52
Duncan, A., 50, 112, 129, 141
Durkheim, É., 1, 11–12, 20–22, 26–27, 29–30, 34–36, 52–55, 57, 96, 123–24, 129, 131–32, 135–38, 140, 143–44, 152, 160, 183

Edwards, A. L., 195
Edwards, C., 18
Efron, R., 7, 78, 109–110
Ehrenwald, J., 75
Eisendrath, C. R., 83
Elder, B., 14
Eliade, M., 21, 26, 28, 32, 43, 133, 151–52
Elias, N., 29, 64–65, 68
Elkin, A. P., 25–26, 31–32, 163–65, 175
Ellis, A. W., 150, 156
Ellis, D. P., 156
Emory, G. R., 97
Evans-Pritchard, E. E., 41

Fabbro, F., 27
Ferrarotti, F. , 64, 113, 167, 188
Fiske, A. P., 3, 8, 95, 102–104, 109, 157, 161, 169, 208, 215
Fiske, J., 13
Fitzpatrick, D., 146
Flaherty, M. G., 72, 188, 190
Foucault, M., 165
Frank, L. K., 167
Franz, M. L. von, 107
Frazer, J. G., 13
Freeman, N. H., 79–80
Friedland, R., 215
Fuster, J. M., 89, 91

Galaburda, A. M., 74–75, 116
Gale, I. F., 30
Gallacher, J. D., 149
Galton, F., 78
Galton, M., 63
Gazzaniga, M. S., 75
Gell, A., 2
Gellner, E., 12, 167–68
Geschwind, N., 74–75, 116
Giddens, A., 172
Gilbert, P., 97
Gillen, F. J., 22, 44, 137
Goffman, I., 165
Gondarra, D., 37
Goody, J, 152
Gosden, C., 57, 170
Gould, S. J., 14–15
Gouldner, A. W., 156
Grayden, W., 14
Greenfield, P., 74
Grey, A., 81, 131, 160
Gudschinsky, S. C., 221n4

Hall, E. T., 63, 65–66, 106, 165
Halpern, F., 112–13, 173
Halstead, W. C., 91
Hampton, K., 150
Handel, B., 74
Harris, J. W., 46, 175
Harris, P., 32, 36, 38–39, 50–51
Harris, Stephen, 148
Harris, Stewart, 173
Hassard, J, 132, 170
Hausfeld, R. G., 146
Heidegger, M., 5, 58–59, 66–68, 86–88, 92–93, 113, 119, 145, 148, 165–66, 170, 187–88, 206
Heine, S., 59, 85–86, 166.
Hellige, J. B., 72
Hermelin, B. M., 79–80
Hertz, R., 27
Hiatt, L. R., 20
Hobbes, T., 16, 156
Hodgson, D., 84
Holm, N., 146–48
Howitt, A. W., 20, 27, 29–30
Hubert, H., 131–32

Name Index

Hughes, R., 14, 16
Hurtig, R. R., 74
Husserl, E., 154–55
Huxley, A., 151

Isambert, F.-A., 132
Isbell, L. A., 98
Itani, J., 2, 99, 107

Jacobs, J. M., 17
James, W., 83–84, 135
Jaynes, J., 215
Johnson, M., 170
Jolly, C., 97
Jones, G. R., 21
Jung, C. G., 211

Kaberry, P., 44
Kalweit, H., 150–51
Kaplan, C. D., 189
Katovich, M. A., 191
Kearney, G., 146
Kearney, M., 76
Keen, I., 37–38, 162–63
Kern, S., 11, 17, 59, 62, 188–89
Kim, H.-J., 85
Kirk, R. L., 18–19
Klatzky, R. L., 77
Klich, L. Z., 80
Kluckhohn, F. R., 86
Kohli, M., 98, 188
Kolig, E., 152, 158, 173
Kunzig, R., 78–79

Lakoff, G., 170
Landauer, T. K., 78
Langer, E., 86
Langer, W. L., 17
Larson, D. W., 156
Laughlin, C. D., 89, 92, 112
LeDoux, J., 100
Lee, D., 64
Leeuw, G. van der, 126–27
Lendon, R., 50
Lestienne, R., 171
Lévi-Strauss, C., 30–31, 127–29, 152
Levy-Agresti, J., 6, 76, 81, 114

Lévy-Bruhl, L., 154
Lewine, J., 77
Lewis, D., 131
Lewis, W., 59
Liberman, K., 54–56, 132, 140–42, 144–45
Lovejoy, F., 149
Luckmann, T., 188
Luhmann, N., 67
Luria, A. R., 6, 9, 72–73, 90–91, 93, 117–18, 215

MacLean, P. D., 100, 102, 109, 157, 215
Maddock, K., 20, 27, 55, 122, 244
Magoun, H. W., 72
Maines, D. R., 191
Malinowski, B., 105, 151
Mann, H., 211
Marshack, A., 61
Marx, K., 172
Masuzawa, T., 33, 70
Mattingley, C., 150
Mauss, M., 11, 132, 151
McConvell, P., 48
McElwain, N., 26
Mead, G. H., 169, 183, 189–90
Meehan, B., 174
Meggitt, M. J., 18, 142, 178
Merleau-Ponty, M., 154
Milner, B., 91
Minkowski, E., 87, 191
Mol, H., 161–62
Montagner, H., 99
Morphy, H., 38
Morris, W., 104
Morton, J., 43–46
Moyer, R., 78
Mulvaney, D. J., 18
Mumford, L., 172
Munn, N., 32–33, 43–44, 46, 123, 138–39, 151
Myers, F., 18, 28, 34, 42–43, 45–46, 53, 122–23, 129, 136, 141, 144, 147–48, 150, 160–62, 176–79, 181–82

Nachson, I., 78
Nass, R., 75
Nettleton, N. C., 74, 76
Newcombe, F., 124
Newton, I., 64, 222n3
Nobel, J., 114–15
Nowotny, H., 67, 172

Oates, L. F., 19
O'Connor, N., 79–80
O'Malley, P., 175
Orton, S. B., 114
Ortony, A., 109
Osmond, H., 211

Passingham, R. E., 90
Payne, R., 97
Phillip, A., 27
Piaget, J., 80
Plateau, J. A. F., 70
Plutchik, R., 7–8, 99–100, 102, 104–106, 109, 121, 137, 157, 169, 215, 222n1
Polich, J. M., 76
Postone, M., 172
Power, M., 98
Pribram, K. H., 6, 73, 88, 90–93, 115, 215
Prigogogine, I., 64

Rabbit, P., 93–94
Radcliffe-Brown, A. R., 22, 152, 176
Rammsayer, T. H., 6
Rappaport, R., 138
Read, P., 18
Reason, D., 172
Reynolds, H., 14, 221n1
Richardson, J., 87–88, 187
Ricoeur, P., 187
Risberg, J., 124
Roberts, R. G., 21
Roget, P. M., 9, 191–93, 196–98
Rosaldo, R., 152
Rose, D., 33–34, 37–39, 42–43, 51–53, 131, 140, 145, 162–63, 174, 177
Rose, F. G. G., 18, 176, 180
Rosebery, L., 17

Rotenberg, R., 65
Rowley, C. D., 182
Rowse, T., 18
Rudder, J., 18, 22–23, 37, 39–40, 46, 50–51, 146
Rutz, H. Z., 165

Sagen, C., 215
Sahlins, M., 136
Sansom, B., 29, 40, 80, 112, 130, 133, 136–37, 143, 45, 153, 160–61, 176, 182, 222n1
Scheler, M., 7–8, 102–103, 109, 157, 215
Schutz, A., 33, 188
Schwartz, B., 63, 101
Seagrim, G., 50
Seyfarth, R. M., 98
Shellshear, J. L., 13
Siegler, M., 211
Slife, B. D., 222n2
Smith, A., 170
Smith, M. A., 21
Smyth, R. B., 13
Sommer, B. A., 221n3
Sommerlad, E., 149
Sousa, R. de, 92
Spencer, W. B., 44, 137
Spengler, O., 66, 167
Sperry, R., 6, 76
Spinnler, H., 124
Springer, S. P., 215
Stambaugh, J., 85
Stanner, W. E. H., 12, 22, 26–27, 30, 53–54, 122–23, 133, 135, 137, 139–40, 149, 157, 165, 173
Stengers, I., 64
Stevens, S. S., 70–71
Strehlow, C., 22, 34, 39, 43–44
Strehlow, T. G. H., 30, 42–43, 48, 147, 181
Strodbeck, F. L., 86
Sturmer, J. R. von, 12–13, 15, 140–41
Stuss, D. T., 91
Sugrue, N. M., 191
Swain, T., 20–23, 26, 36–37, 39, 50, 53, 181, 214

Talpin, G., 14
Tatz, C., 17
TenHouten, W. D., 7, 101–102, 108–109, 124–26, 131, 197, 203–206, 217–19
Terdiman, R., 215
Thompson, E. P., 171–72, 200, 202, 208
Thomson, D. F., 38
Tonkinson, R., 18, 144, 152–53
Tönnies, F., 7, 96, 169
Traweek, S., 23, 32, 52
Treisman, M., 6
Tribolet, N. de, 78
Trivers, R. L., 156
Tubbs, W. E., 89
Turner, D. H., 122, 178
Turner, F., 81, 114
Turner, T. J., 109
Turner, T. S., 127–28

Vernon, P. E., 166
Vogel, W. H., 6

Wake, C. S., 15
Warden, S. A., 166
Warm, J. S., 70–71

Warner, W. L., 18, 37–38
Weber, M., 172, 209
Weil, J., 79
Welsby, T., 27
Whatmore, G., 98
White, J. P., 18
Whitehead, A. N., 83–84, 135
Whitrow, G. J., 1, 47, 60, 206, 215, 222n3
Whorf, B. L., 52
Williams, N., 38
Williams, R., 136
Woods, J. D., 14
Woodburn, J., 176
Woodcock, G., 67
Wurm, S. A., 19
Wynn, K., 79

Yallop, C., 20, 49–50, 221nn3–4
Young, A. W., 77
Young, M. D., 131, 168

Zelazo, P. D., 93
Zern, D., 166
Zerubavel, E., 60, 65, 210
Zimmerman, M. E., 87
Znaniecki, F., 7

Subject Index

abidingness, 31, 37, 39, 54, 133, 135, 138, 161–62, 181
Aborigines of Australia: languages, language use, 13, 18–20, 22, 38, 47, 55, 130, 143, 164, 221nn3–4; law, 31, 36–37, 56, 133, 138, 148, 159, 162, 164–65, 175, 180, 214; estimates of population, 14, 18; religion (*see* Dreaming, the)
Aboriginal Australian peoples or "tribes:" Anindilyakwa, 46–47; Aranda, 20, 34–35, 39, 43–44, 48, 50, 137; Jankudjara, 178; Kuuk-Thayorre, 47; Mardudjara, 152; Nunugal, 27; Pitjantjatjara, 43, 138, 178; Pintupi, 28, 43, 45, 53, 122–23, 129, 141, 150, 158, 160–62, 176–82; Walbiri, 23, 32, 43, 52, 138, 181; Waringari, 178; Warmun, 180; Yarralin, 42, 177; Yolngu (Murngun), 22–23, 37–39, 50–51, 139, 162–63
acceptance: and identity, 56, 101, 105, 133, 137–38, 155, 189; and time, 54, 62–63, 85, 87, 137–38, 155, 189; as primary emotion, 101, 137–38
age: in years, 37–38, 50, 56, 67–68, 80, 125, 163, 174, 191–92, 199, 205; qualitative grouping, 38, 41, 44, 46, 50, 56, 162–63, 174
agency, 4, 55, 103, 107, 151. *See also* communion
aggression, 97–99
agonic society, 5, 8, 95–100, 107–109, 113, 119, 184–86, 198, 200–202, 208–209, 211–15, 219

analysis, 5, 74, 76–77, 91,113, 116–17, 211. *See also* logical–analytic information processing; holistic thought
ancestors, 11, 21, 23, 29, 31–35, 37, 39, 42–44, 48, 52–53, 105, 123–24, 126, 139, 146, 163, 174, 177, 179
anger, 101, 140, 157
animism, 17–21, 28, 35, 163
anticipation, 15, 53, 60, 65, 85–86, 88–89, 91–92, 97–98, 102, 119, 140–41, 145, 148, 166, 168–69, 188, 212
apes. *See under* primates
appositional informational processing. *See* gestalt–synthetic information processing
arithmetic operations. *See under* number
assimilation, 14, 22, 149–50, 173, 209, 215
asymmetry, 73–74, 77, 105
attention, 38, 86, 90–91, 97–99, 116, 131–32, 140, 142, 148, 154, 161, 164
auditory processing. *See* hearing
Australia, 12–21, 51, 147–50, 173, 176, 182–83, 195, 209, 212, 214
Australian Aborigines. *See* Aborigines, Australian
Australians. *See* Euro–Australians; Aborigines of Australian
authority-ranking social relations: and episodic-futural time, 4, 10, 157–68, 185, 207–208, *219*; as component of agonic society, 97, 184–86, 202, 212–13; as opposite of equality–matching, 4, 95, 107–108,

251

252   Subject Index

160, 185; definition and measurement of, 8, 101, 105–106, 108, 157, 186, 196–97, *217–18*; in Aboriginal culture, 137, 153, 156–65, 212, 214; in modern society, 156, 165–68, 171, 214

being, 7, 28, 34–35, 44, 50, 85–87, 102, 107, 127, 139, 144, 146, 162
birth: in Aboriginal culture, 18, 21–22, 30; in modern society, 38, 67–68, 188; and communal sharing, 104, 121, 198. *See also* reproduction
boredom and passage of time, 71–72
boundaries, 18, 64, 102, 106, 129, 151, 154, 162
brain, functional units of, 9, 72–73
brain, structures of: amygdala, 100; angular gyrus, 78; association areas, 73, 89, 91; basil ganglia, 79; cerebral hemispheres: left hemisphere, 6–7, 57, 69, 73, 81, 83, 89, 91, 100, 113–14, 116–18, 192; right hemisphere, 6–7, 69, 73–81, 83, 89, 93, 100, 113–14, 116–17, 125, 215; cingulated gyrus, 91; corpus callosum, 79, 81, 114; hippocampus, 98, 100; hypothalamus, 100; limbic system, 6, 89–91, 93, 100; lobes of the cortex: frontal, 6, 73, 81, 89, 91, 112, 115, 117–18, 215; occipital, 6, 72, 89; parietal, 6, 72, 78–79, 89; temporal, 6, 72, 89; precentral gyrus, 72; posterior cortex, 6, 9, 72–73, 84, 89–91, 112, 115–19, 215; reticular activating formation, 72, 90; speech areas: Broca's area, 74, 118; Wernicke's area, 78, 114

calendars, 3, 5, 9, 11, 26, 34, 36, 48–49, 57, 61–63, 65, 67–68, 70, 88, 106, 113, 172, 185, 192, 214, 222
capitalism, 1, 3, 60, 67, 103, 112, 170–73, 176
categories of the understanding, 1, 12, 85, 192–93
causes and effects, 4, 9, 31, 163, 192, 197–202
centuries, 34, 61

cerebral commissurotomy (split brain surgery), 79, 114
cerebral hemispheres. *See under* brain, structures of
cerebral lateralization, dual brain theory, 6–7, 9, 69, 72–81, 89, 91, 110, 215. *See also under* brain, structures of (left hemisphere, right hemisphere)
ceremony. *See* ritual
change, 1, 15, 30, 35, 39, 48, 54, 62, 64–66, 70, 84–86, 90, 97, 128, 131–32, 150–52, 162, 167–70, 175, 222
chimpanzees. *See under* primates
Christianity, 22, 36–37, 151, 162, 170, 175
chronology, 48, 64, 66–68, 188, 221n4. *See also under* time, measurement of
clans, 18, 21, 34, 48–49, 53, 121–24, 135, 160, 176, 181
classification, categorization, 1, 7, 9–11, 20–21, 28, 38, 40, 48, 59, 74, 85, 90, 99, 102, 122, 128–29, 131, 178, 182, 191–93, 196, 209–210
clever fellows. *See under* sorcery
clocks, clock time, 3, 5–6, 9, 14, 25, 38–39, 47–49, 56–58, 61–68, 70, 83, 106, 113, 158, 166, 171–72, 183, 185, 192, 214
closure, visual. *See* gestalt–synthetic information processing
cognition, 2, 4–6, 20, 50, 60, 63, 69, 74–75, 77, 80, 89, 91–93, 95, 113, 116, 119, 124, 129, 150, 184, 189, 191–93, 196, 207, 209–211, 214–15, 222
collective effervescence, collective sentiments, 54–55, 104, 132, 137, 143
colonialism, 8, 16–17, 19, 22, 178
command, control, 6, 69, 88–94, 97, 102, 106, 149, 161–62, 187, 190
commodification of time, 67, 170–172
commodities, 62, 106–107, 151, 169–72, 176, 180

Subject Index

communal-sharing social relations: and patterned-cyclical time, 121–56, 185, 199, 207, 219; as component of hedonic society, 2–4, 108, 184–85, 212; as opposite of market–pricing social relations, 4, 8, 95, 107, 185; defintion and measurement of, 104–05, 186, 196, 217–18; distinguished from equality–matching, 103; in Aboriginal culture, 121–56, 159–160, 184, 212; in modern society, 131

communication, 27, 63, 74–75, 84, 99, 101, 116, 131, 141, 145, 149, 164–65, 187, 189

communion, 4, 107, 144. *See also* agency

community, 4, 7–8, 18, 20, 37, 55, 80, 96, 98–100, 102–103, 108, 119, 121–22, 129, 131–32, 136, 138–42, 144, 147, 149, 151, 155, 158–60, 163, 173, 178–79, 184, 186, 189, 202, 208–209, 212, 214, 221

competition, 2, 13, 92, 103, 107, 112, 151, 172–73; for time, 171

complementarity, 86, 96, 107–109, 111–13, 116–19, 212, 215. *See also* opposition

complexity, 5, 9, 20, 52, 61, 74, 81, 89, 92, 102, 106, 117, 123–24, 171, 189, 211

conation. *See* reason

conception. *See under* reproduction, sexual

conditional equality, 4, 8, 99, 107–108, 135, 145, 212. *See also* identity; equality-matching

conformity, 105, 140

consciousness, 7, 63–64, 75, 83–84, 92, 110, 119, 128, 132, 137, 165, 175, 190

consensus, 55–56, 101, 105, 112, 132–33, 136, 139–45, 159–60, 163. *See also* likemindedness

content analysis, 9–10, 191–93, 187, 209–210

continuity, 22–23, 26, 33–35, 44, 48, 52–53, 57–58, 60, 64–65, 75, 83–86, 90, 97, 122, 128–32, 144, 148, 151, 159, 162, 166, 176, 182, 211, 222. *See also* discontinuity

contracts, 16, 64–65, 177

control and command: and executive mental functions, 6, 59, 88, 97, 117, 119, 166, 170–72; and social rank, 102, 106, 171; and time, 59, 63–64, 113, 161, 165, 172, 187, 196; in Aboriginal culture, 14, 27, 31, 129, 162, 172

corroborees. See dancing

country, 15, 28, 32, 41–43, 46, 131, 141, 148, 158, 163, 169, 176–78, 181–82, 221n1. *See also* land; place

culture, 1, 5, 7, 20–25, 32, 36, 48, 53, 56, 60–63, 65, 70, 75, 80, 121–22, 135–36, 141, 146–49, 149, 152–55

cyclical time. *See* patterned, cyclical time

dancing, 30, 32, 144, 161, 180

Darwin, Northern Territory, 40, 80, 112, 136, 143, 153, 160–61

Dasein, 87, 166

dateability, 66, 113

days, 12, 26, 37–40, 46, 52, 60–63, 65–68, 131, 148, 158, 165, 170–71, 174, 213–14, 222

deadlines, 64, 113

death, 21, 31–32, 38, 40–41, 43, 46, 48, 53, 57, 67, 87, 100, 121, 140, 145, 148, 163–65, 181–82, 188

decades, 61, 63, 205–206, 209

decision–making, 90, 105, 132, 136, 140, 142–43, 155, 159, 161, 163, 191, 214

discontinuity, 26, 34–36, 69, 74, 128–29, 145. *See also* continuity

disgust, 101, 137–38

dominance, social dimension, 97–99, 101, 109, 112, 146–47, 150, 176, 210

double polarity. *See* polarity; quaternio

Dreaming, the, 22–23, 26–28, 30–34, 41–45, 51–53, 123, 127, 131–32, 135, 137–39, 144–47, 150, 153, 159, 161–63, 165, 178, 182

duality: of time, 2–3, 8, 26, 29, 32, 52, 69, 123–24, 127–29, 152, 193, 210; of the sacred and the profane, 21, 26–30, 137; of social organization, 7, 96, 100, 109, 162; social duality theory, 2, 7–8
duration estimation, 6, 52, 70–71
duration. *See* long duration

economy, 2–4, 7, 14, 18, 21, 62, 65, 67, 102–103, 112–13, 116, 142–43, 149, 151, 159, 167–74, 181–84, 201, 207–208, 212, 214. *See also* market pricing
education, 14, 22, 38, 51, 63, 65, 67, 80, 112, 116, 148–49, 161, 166, 183
effervescence. *See* collective effervescence
elders, 29, 43, 56, 136, 158–59, 163, 174, 180–81, 214
embarrassment, 141
emotions: primary and secondary adaptive reactions, 7–8, 60, 99–102, 109, 137, 157, 169, 222; and ritual interactions, 54–55, 129, 137–38, 142, 144, 163–64, 190; and episodic–futural time, 87–88, 91–92; and right cerebral hemisphere, 74, 93, 116. *See also* psychoevolutionary theory of emotions; collective effervescence
environment, 13–15, 18, 29, 41, 54, 60, 67, 75, 80, 91–92, 123, 138, 145, 147, 154–55, 190
episodic information processing, 6, 9, 73, 88–94, 112, 115, 117, 185, 196, 215
episodic-futural time: and authority-ranking social relations, 95, 157–68, 185, 200, 207, *219*; and the brain, 6, 73, 89–93, 116–17; and the emotions, 87–88, 91–93; complementary to linear time, 5, 9–10, 113, 117–18, 185, 192, 196–97; definition and measurement of, 5, 88, 90, 116, 192, 217; in Aboriginal culture, 157–65; in modern society, 165–68, 184, 204; opposite of immediate-participatory time, 9, 86, 89, 92–93, 111, 115–16

equality-matching social relations: and identity, 105; and immediate–participatory time, 9, 99, 135–56, 185, 207, *219*; as component of hedonic society, 4, 108, 184, 186; definition and measurement of, 99, 103, 105, 196, 198, *217–18*; complementary to communal sharing, 9, 160; in Aboriginal culture, 129, 160, 177, 212; in modern society, 146–47, 155–56, 214; opposite of authority ranking, 8, 95, 107, 111–12. *See also* conditional equality
equilibrium, 90, 97, 135, 164
eternity, 22, 42, 52, 145, 182
Euro-Australians, 18, 50–51, 125, 146–47, 173, 184, 195, 198, 200–202, 205–206, 209, 212
Europeans, 13, 15, 17–19, 31, 38, 67, 146, 148, 167, 172–73, 195, 222n2
events, 23, 26–28, 31–32, 36–40, 42, 47, 51–53, 57, 60–68, 73, 81, 84, 89, 92, 112, 126–31, 133, 136, 139, 142–43, 145–46, 148, 154, 158–59, 161–63, 167, 172, 180–81, 187–88, 191. *See also* happenings
evolution, 2, 5, 7, 12–15, 17, 73–74, 90–91, 96, 99–103, 116, 119, 156, 210
exchange, 7, 48, 50, 67, 96, 105–107, 123, 139–41, 144, 151–53, 156, 169–70, 172–74, 176–78, 180
executive mental functions, 88, 93–94. *See also* episodic information processing
existence. *See* being
expectation. *See* anticipation
exploitation. *See under* economy

family, 18, 41, 49, 80, 96, 104, 121–23, 129, 139, 144, 149, 158, 160, 173–76, 183, 201, 209, 212, 214
fear, 29, 101, 157, 163
figure, as opposed to ground, 60
finitude, 84–85, 87, 100, 147, 170–71, 188. *See also* temporality

formal operations, 80
fourfoldness: of elementary social relations, 4, 7–8, 95, 111, 119, 161, 184, 198, 208, 215; of time consciousness, 2, 4–6, 9–10, 95, 111, 119, 185, 191–92, 196, 207, 211, 215; of modes of information processing, 1, 3, 111, 119, 184, 207, 211, 215; of problems of life, 7–8, 95, 215, 222
future, the, 18–19, 36, 40, 80, 111–13, 115–17, 119, 127, 130, 132, 136, 145–50, 157–59, 161, 165–68, 170–71, 173, 177, 182, 187–92, 197, 202, 205, 209, 211, 214, *217–19*. *See also* episodic-futural time

gathering. *See* hunting and gathering
Gemeinschaft, 7, 96
generations, 39, 46, 48–49, 52, 64, 131–32, 145, 163, 177, 221
genocide, 14
Gesellschaft, 7, 96
gestalt-synthetic information processing, 2, 7, 30, 36, 41, 57, 59, 69, 75–77, 80–81, 89, 93, 111–12, 116, 123–25, 128, 131, 160, 185–88, 211
ghosts, 21, 182
grief. *See* sadness
ground, as opposed to figure, 28, 60

handedness, 6, 27, 73, 78, 114, 117, 164
happenings, 30, 36–40, 61, 112, 116, 130–31, 133, 143, 153, 158. *See also* event
happiness. *See* joy
hearing, 32, 70–71, 83, 89, 133, 189. *See also* sound
hedonic society, 4, 8–9, 95–96, 98–100, 107–109, 119, 160, 184–86, 198, 201–202, 208–209, 211–12, 214–15, 218–*19*
hierarchy, social: and authority ranking, 4, 105–108, 157, 161; and social organization, 7, 63, 96–99, 101–102, 106, 157, 160–61, 171, 176, 185; and time, 81; as problem of life, 7–8, 95, 97, 100, 101–108, 157; in the brain, 74, 84; opposite of conditional equality, 107–108, 160
history, 5, 7, 63–64, 66–67, 70, 103, 106, 116, 119, 188–90, 210–11, 215: in Aboriginal society, 11, 14–16, 28, 30–31, 42–43, 54, 132–33, 137, 152; in western civilization, 12, 36, 165, 167, 172, 215
holistic reason, 2, 25, 76–77, 85. *See also* synthesis; gestalt–synthetic information processing
Hopi, Native American people, 52
hours, 14, 50, 61, 63, 66, 112, 170
hunting and gathering, 2, 18, 21, 42, 47, 80, 101, 121, 148–49, 158, 160, 174, 176, 178, 212–14
husbands, 44, 131

ideas, 9, 55–56, 83–84, 93, 117–18, 132, 140–43, 147, 167, 179, 192
identity: and emotions, 137; and equality matching, 105, 155; and social organization, 7–8, 103, 106–107, 155; as problem of life, 7–8, 95, 99–103; in Aboriginal culture, 21, 53, 136, 138, 148, 150–51, 161–63, 176–77, 179, 181–82, 198
images, imagery, 34, 75, 78–79, 90, 93, 114, 123, 128, 190
immediacy. *See* moment
immediate-participatory time: and equality-matching social relations, 5, 95, 119, 135–55, 188, 200, 207, 219; and the brain, 73, 76, 83, 89; and the future, 187, 190; component of natural time, 9, 184–85; complementary to patterned–cyclical time, 112, 160; definition and measurement of, 83–86, 145, 154, 217; in Aboriginal culture, 28, 32, 53, 155, 157, 161–62, 174; opposite of episodic-futural time, 5, 9, 83, 86, 89, 93, 111, 115
immortality, 53
impermanence, 33, 85
inapatua, 35, 44

industrial society, 3, 60–61, 63–65, 67, 170–72, 176, 184, 206
inequality, 8–9, 99, 101, 106, 108, 147. *See also* conditional equality
initiation, 27, 29–31, 35, 37, 43, 45, 55, 123, 127, 138, 147, 155, 163, 177, 180–81
intelligence, 1, 13–15, 17, 54, 91, 124, 210
intentionality, 87–92, 112, 118, 129, 167

jealousy, 129, 174, 179
joking, 38–39, 159
journeys, 41–42, 153, 155, 158
joy, 4, 48, 55, 59, 101, 104, 106, 152

kinship, 48–51, 104–105, 122–24, 129, 158, 160, 178
knowledge, 29, 37–38, 42–43, 56, 63–64, 98, 127, 131, 138, 143, 145, 153, 159, 162–63, 166, 170, 177, 179–80, 188–89, 210

labor, 9, 11, 14, 65, 67, 105–106, 143, 146, 149, 153, 155, 165–66, 170–76, 183, 200, 208
land, landscape, 12, 14, 17–18, 21–22, 27–28, 32, 34, 39, 41–44, 53, 105–106, 122, 124, 138, 147–48, 150, 153–55, 157, 161–62, 169–70, 173–79, 181, 183, 213–15, 221n1. *See also* country; place
language, 25, 52, 130, 187, 189, 191–92; and the brain, 73–79, 116–18, 192. *See also under* Aborigines of Australia
law, 13, 16, 22, 33, 37, 56, 133, 145, 148–49, 159, 161–62, 164–65, 175, 180, 214. *See also under* Aborigines of Australia
left hand and side of body, 27, 76, 78–79, 114, 125, 164
life community. *See* community
life cycle: and reincarnation, 21, 33, 37, 41–44, 48–53, 131, 164; stages of, 41, 95, 100, 116

life–histories, life–historical interviews, 146, 183, 186, 188–90, 213
likemindedness, 130, 135–36, 143–44, 160, 185
linear time. *See* ordinary–linear time
local time, 62–63
logical–analytic information processing, 6–7, 15, 57, 69, 73–74, 76–77, 81, 89, 91, 111, 113, 116, 185, 192
long duration, 26, 41, 48, 51–56, 66–68, 87, 132–33

Maori, 27, 221
market-pricing social relations: and ordinary-linear time, 4, 119, 169–86, 200, 207, *219*; as component of agonic society, 3–4, 108, 185, 200, 202, 211–13; as opposite of communal-sharing, 4, 8, 95, 107; definition and measurement of, 8, 102–104, 106–107, 169, 195–96, 214, 217–18; in Aboriginal culture, 173–84, 212, 214; in modern society, 51, 169–72, 212
marriage, 27, 38, 103, 123, 139, 159, 177, 198
memory, 32, 37, 39, 79, 84, 86–87, 91, 166–67, 187–88, 190–91
men, 29, 31, 42, 123,129, 141, 150–51, 153, 162–63, 165, 174–75, 213
mind, 5, 8, 12–14, 36, 54, 73–75, 83–89, 107, 116, 119, 132, 137, 144, 152, 164, 166, 171, 184, 190–92, 209–211, 214–15
mindfulness, 6, 83–86, 116, 155, 158, 191. *See also* moment; now; present, the
mindlessness, 86
minute, 60–61, 63, 66, 112, 183
moment, 6, 29, 32, 40, 51, 55–57, 60, 63, 77, 83–86, 88, 90, 98–99, 141, 143, 145, 151–52, 167, 188, 190–91. *See also* now; present, the; mindfulness
money, 67, 106, 149, 169, 176, 212, 214; and time, 9, 60, 67, 170, 172, 176, 183
monkeys. *See* primates

Subject Index

month, 12, 30, 47, 61, 67, 131, 170, 222
myth, 21–22, 27, 30–32, 34, 38, 40, 42–43, 123–24, 126–28, 133, 139, 150, 152, 163, 173, 177, 181

narratives. *See* stories
natural time experience, 5–7, 9, 96, 112–115, 119, 133, 160, 184–86, 198, 201–202, 205–206, 208–209, 212, 214–15, 219
natural will, 7, 96. *See also* rational will
neomammalian brain. *See* triune brain
neurocognitive hierarchical categorization analysis. *See* content analysis
night, 40, 46
noble savage syndrome, 17
now, 17, 25, 31, 33–34, 37, 58–59, 61–62, 66, 68, 85–87, 133, 145, 166, 170, 189. *See also* present; moment; mindfulness
Nuer, 42
number: and time, 36, 50–51, 57–58, 63, 66–68, 76–79, 109; addition, 103–104, 106–107; division, 12, 26, 63, 66–68, 89, 170, 174; multiplication, 44, 57, 79, 188; numerical cognition, 57, 67, 192; subtraction, 100, 104, 107. *See also* fourfoldness

object, 73, 75, 79, 84–85, 87, 92. *See also* subject
opposition, opposites, 2–6, 8–9, 25, 44–45, 57, 59, 68, 86, 95, 101, 103, 107–109, 111–16, 137, 182, 199. *See also* complementarity
oral history. *See* life historical interviews
ordinary-linear time: and market pricing social relations, 4, 95, 106, 119, 169–85, 200, 207–208, 219; and the brain, 6–7, 73–79, 83, 113, 118; complementary to futural orientation, 4, 5, 9–10, 67, 113, 118, 185, 202, 209, 212; definition and measurement of, 2, 5, 25, 52, 57–59, 65–66, 70–73, 192, 197, *217*; in Aboriginal culture, 152, 173–83, 200, 202, 205; in modern society, 2, 26, 52, 59, 67, 69–71, 169–73, 202, 205, 209; opposite of patterned–cyclical time, 2–3, 8–9, 25, 57, 86, 88, 222

paleomammalian brain. *See* triune brain, the
parallel information processing, 76–77
Paris International Conference on time, 62
participatory information processing, 6, 9, 73, 85–86, 88–90, 92–93, 99, 112, 115–17, 135, 154–55, 185, 215, 219
participation, phenomenological, 154
past, the, 2, 12, 14–15, 23, 25–26, 30–31, 34, 37–38, 40, 45–46, 51–52, 57–59, 66, 68, 84–88, 92, 111–12, 116, 121, 124–28, 131–33, 139, 143, 145–47, 151, 158, 166–68, 171, 177, 182, 187–91, 193, 197, 211, 221
past, the, and the present, 2, 23, 26, 30–32, 34, 51, 84–87, 92, 124–28, 132–33, 139, 146, 166–67, 182, 187, 189–90, 193, 197, 211
patterned–cyclical time: and communal–sharing social relations, 121–26, 128–33, 199, 219; and the brain, 7, 36, 75–81, 185, 211; complementary to present orientation, 9, 160, 184–85, 208; definition and measurement of, 3, 5, 25, 86, 111, 193, 217; in Aboriginal culture, 25–56, 123, 207; opposite of ordinary–linear time, 3, 5, 9, 25, 68, 111
perception. *See* sense perception
permanence, 23, 32, 51, 53, 70, 85, 122, 132, 145, 165
place, 12–13, 21–23, 28, 30–33, 37, 39, 44, 46–47, 53, 102, 106, 122–24, 128, 143, 159, 163, 169, 174, 177, 179–81, 222n1. *See also* country; land

planning: in Aboriginal culture, 27, 37, 146, 158–59, 162–63; in modern society, 63–64, 113, 146, 165–67, 188–89; of the brain, 6, 73, 78, 88, 94, 119
Plateau's Law, 70
politics, 3–4, 7, 55, 102–103, 106, 140, 147, 157, 159, 175, 177, 181, 212, 214. *See also* authority–ranking social relations
posterior cortex. *See under* brain, structures of
posteriority, temporal, 57, 65, 105
power, social. *See* hierarchy
prefrontosensorial polarity principle, 92, 112
prejudice, 18, 147
Preliterate society. *See* primitive society
present, the, 1–6, 17, 23, 25–26, 30, 33, 37–38, 40, 51–53, 57–59, 68–69, 75, 83–86, 88, 111–12, 116, 125, 131, 135, 145–48, 150, 158, 166–67, 171, 187, 197, 202, 205, 207–209, 211. *See also* immediate–participatory time; moment; now; mindfulness
Present and past. *See* past and present
primary emotions, 8, 60, 100–102, 109, 137, 157, 169, 211, 222. *See also* specific emotions; emotion
primates: apes, 74, 91, 98, 119, 184; monkeys, 75, 96–98, 114, 119, 184
primitive society, 2–4, 11–12, 21, 26, 43, 103, 105, 151–52, 210
priority, temporal, 57, 157
problems of life, fundamental. *See under* identity; temporality; hierarchy; territoriality
profane, the, 11, 21, 23, 26–27, 29–30, 36, 124, 128–29, 137, 152
progress, 12–13, 60, 133, 152, 166–68, 209
propositional information processing. *See* logical–analytic information processing
propositions of present theory, 4, 9, 119, 121–22, 184–86, 193, 199–202, 208

psychoevolutionary theory of emotions, 7, 99–102
psychophysiology of duration estimation, 70–72

quaternio the social, 8, 107–109, 111. *See also* opposition; complementarity

race, 13, 15, 17, 22, 104, 175
rank, social. *See* hierarchy
rational time experience, 5, 7, 9, 96, 111, 113, 117–19, 184–86, 198, 202, 205–206, 208–209, 214–15, 219
rational will, 7, 96. *See also* natural will
rationality. *See* reason
R–complex. *See* reptilian brain; triune brain
reaction time, 76–77
reason, rationality, 1, 3, 12, 50, 69, 74–75, 79, 87, 89–94, 96, 103, 106, 109, 111, 117, 123, 146, 170, 191–92, 209
reciprocity, social, 105, 138–39, 144, 151–53, 156, 185
reincarnation. *See under* life cycle
religion, 11, 14–15, 20–21, 26–30, 32, 37, 43–44, 58, 103, 131, 137–38, 143–44, 147, 150–53, 170, 173, 179, 214. *See also* Dreaming, the
reproduction: cultural, 28, 43, 56, 112, 123, 136, 179; sexual, 29, 44, 56, 95, 99–102, 104, 121, 125, 188, 198
reptilian brain (R–complex), 100; *See also* triune brain, the
resources, 9, 14, 48, 63–64, 97, 104–106, 153, 169–70, 173, 178, 185, 189, 209; and time, 67, 116, 170, 173, 185, 189
respect. *See* acceptance
reverted escape, 97–98
rhythmic events, 11, 23, 39, 118, 132, 162. *See also* events
right hand and side of body, 6, 27, 73, 76, 78–79, 114, 117

Subject Index

ritual, 11–12, 18, 20, 22–23, 26–32, 36–39, 42, 46, 50, 54–55, 103, 121–23, 125–30, 132–33, 137, 140, 143–44, 146, 152–53, 158–59, 161, 163, 176–82

sacred, the, 11, 21, 23, 26–34, 36–37, 39, 42, 52, 64, 122, 124, 126–29, 133, 135, 137–38, 148, 151–52, 159, 163, 177, 179–81. *See also under* duality
sadness, 101, 104, 182
schedules, 3, 5, 49, 63–66, 113, 165, 167, 172, 183, 185, 192
schooling. *See* education
seasons, 11, 39, 42, 46–48, 51, 158, 191
seconds and fractions of seconds, 61, 63, 84, 172
self, 21, 29, 37, 65, 92–93, 98–99, 103, 107, 114, 116, 129–30, 132, 138, 140–42, 147, 151, 154, 156, 166, 191
sense perception, 6, 9, 75, 83, 81–86, 89, 116–18, 155, 192; of Aborigines, 28, 46, 154–55
sequence, 33, 52–53, 57–58, 61–64, 68, 74, 76–79, 81, 87, 94, 105, 128, 166, 172, 188
serial information processing, 7, 56, 76–77, 80. *See also* simultaneous information processing
sex differences in time consciousness, 162, 198–99, 202–206, 219
shame, 129, 141
sight. *See* vision
silence, 55, 83
simultaneity, 5, 7, 32, 52, 57, 65, 68, 76–77, 80–81, 86, 111, 113, 116, 131, 146, 185, 222n3
simultaneous information processing, 7, 80–81, 86, 111, 113, 116, 131. *See also* serial information processing
singing, 27, 30, 33, 38, 42, 126, 144, 152, 154, 161, 163–64, 174, 179–80
smell, 70, 84
social class, 16, 18, 80, 146, 148, 157, 161, 166, 200, 210
social duality theory. *See under* duality

social relations, elementary. *See* equality matching; communal sharing; authority ranking; market pricing
solidarity, social, 54, 64, 96, 104, 143–45, 155, 185
sorcery, 27, 136, 139, 146, 150, 154, 162–65
soul, the, 26, 34, 43–44, 53
sound, 27, 71, 77, 84, 89, 118, 155, 182
space, including spatial reasoning, 1, 23, 28, 32–33, 39, 46, 51–53, 62, 64, 66, 75, 89, 106, 116, 139, 141–42, 147–48, 151, 154, 162, 164, 166, 169, 172–73, 191; and time (*see under* time)
specific emotions: *See* acceptance, aggression, anger, anticipation, boredom, disgust, embarrassment, fear, humility, jealousy, joy, sadness, shame, surprise
speech. *See* language
split brain surgery. *See* cerebral commissurotomy
stories: as measures of time and social relations, 187, 190–91, 198, 213–14; in Aboriginal life and religion, 28, 32–33, 39, 43, 63, 112, 126, 144, 161–62, 179. *See also* content analysis; life histories
subject, 85, 87, 92, 139, 154. *See also* object
suburban. *See under* urban
surprise, 60, 102, 106, 169, 222
survival, 41–42, 48, 53, 70, 80, 100, 106, 116, 123, 136, 145, 175, 216. *See also* evolution
symmetry, 54, 73, 77, 87, 153
synthesis, 36, 80, 91, 93, 113, 124, 127, 131–32, 143, 188. *See also* gestalt-synthetic information processing

tactile sense. *See* touch
taste, 89
technology, 2, 121, 148, 171
temporal compression, 5, 25, 70, 85–86, 111, 116, 145, 185, 190

temporal stretch, 5, 25, 86–88, 90, 111, 113, 116, 148, 174, 177, 185, 187–88

temporality, 2, 4, 6, 15, 26, 32–33: and emotions, 87, 104, 113, 185; as limitation of time, 58–59, 87, 121; as problem of life, 7–8, 99–100, 105, 121; in Heidegger, 56, 86–88, 113, 165–66, 187

temporospatial representation, 81

tense and time, 59, 189

terra nullius, empty territory, 18, 221n1

territoriality: and emotions, 7–8, 95, 100, 102, 169; and hierarchy, 97, 100; and market pricing, 106, 108, 185, 214; as problem of life, 95, 97–99, 101–102, 169, 222; in Aboriginal culture, 21, 44, 157, 176, 178, 183, 214. *See also* land; country

thickness of the present, 66, 84

thought. *See* cognition

time: absolute, 32, 51, 59, 64, 191; abstract, 38, 61, 63, 66–67, 73, 170, 172, 191; and space, 1, 23, 28, 32–33, 39, 52–53, 62, 64, 66, 147, 162, 172–73; being time (*uji*), 85–86; deconstructed, 69; diachronic and synchronic, 48, 65–66, 93, 127–28, 144, 152, 155; discontinuity of, 26, 34–36, 69, 74, 128–29, 145, 211; Greenwich mean time, 62; heterogeneity of, 26, 34–36, 128; homogeneity of, 26, 60–62, 170; horizon of time, 68, 86–88, 190; in general, 132; measurement of (*see* seconds; minutes; hours; days; years; decades; centuries; clocks; calendars; schedules; timetables); oecological, 41; private, 11, 184; public time, 61–62, 65–66, 188–89; qualitative, 5, 26, 36, 41, 50–51, 53, 56, 66, 132, 172; quantitative, 50, 56–57, 61, 63, 66–68, 170; singularization of, 36–37, 126; specific kinds of (*see* episodic-futural; immediate-participatory; patterned-cyclical; ordinary-linear); standardized, 59, 61–62, 66–68, 171–72; transience of, 90; two kinds of emergent time (*see* natural time experience; rational time experience)

timetables, 3, 61–63, 65, 78, 172

totality, 52, 59, 84–85, 127, 191

totemism, 12, 21, 28, 31, 34, 38–39, 42, 44, 48, 124, 128, 135–38, 146–47, 150, 152–54, 214

touch, 71, 74, 84, 98, 137

tradition, 17, 23, 26, 30–31, 38–39, 50–51, 54, 56, 80, 121–23, 125, 128–29, 132, 138, 143, 146–49, 158–59, 164, 169, 173–76, 183–84, 191, 212–14

tribes. *See under* Aboriginal Australian peoples

triune brain, 99–100

Truk, Southwest Pacific, 65–66

urban, urbanization, 3, 14, 18, 60, 65, 79–80, 122, 124, 146, 160–61, 163–64, 170, 172, 174, 183, 199, 206; suburban, 14, 18, 122, 160, 183, 199, 206

validity, 83, 93, 109–110, 197–98, 209

value, values, 7–8, 31, 50, 54–55, 59, 67, 92–93, 97, 102–103, 105–106, 112, 127, 130, 136, 143, 146, 148–50, 152, 159, 166–67, 170–71, 179–80, 185, 211

verbal stimuli, 73, 76–77, 79. *See also* language

vision, sight, 18, 46, 70, 74–80, 84, 89, 91, 108, 114, 124, 182

visions, 11, 46, 70

walkabouts. *See* journeys

watches, 38, 62, 66–67, 171, 183

water clocks, 61

weeks, 12, 61, 165, 183

Western culture and civilization, 2–3, 13–15, 23, 26, 32, 36–39, 50–53, 56, 64, 67, 79–80, 95, 103, 131, 146–47, 150–51, 161, 165–73, 178, 182, 184, 210, 212, 214

will power, 91, 94, 106, 192, 196, *217*

wives, 44, 54, 130–31, 136, 143, 149, 175, 180
women, 18, 27, 32–33, 44, 48, 140–41, 153, 158, 161, 174, 176, 183, 211n2
words, 13, 20, 29, 74, 77, 79, 104, 112, 114, 117–18, 124, 129–30, 141, 143–44, 160 as content–analysis measures, 1, 9–10, 191–93, 195–97, 209–210, 213, *217–18*

work. *See* labor
world view, 21, 75, 138, 151

years, 12, 20, 34, 36, 38, 42, 46, 48, 50–51, 61, 67–68, 124, 131, 148, 167, 174, 181–82, 191–92, 199, 205, 222n2